North Sea Oil – the future life-blood of British industry. It is predicted that by 1980 Britain will be producing enough oil to satisfy her own needs, perhaps even enough to export abroad. Will the dream come true? If it does, her economic future may be assured.

But there are many problems associated with finding the oil, setting up rigs and pumping it ashore. The North Sea is the most difficult area in the world for searching and drilling. Gale-force winds and high seas prevent activity in the winter months. Time lost is money wasted. On shore, where the oil-supply industry is making its mark, there is opposition from the local people. The benefits of the oil must be weighed against its effect on the countryside and the threat to the environment.

Despite the optimistic forecasts, will Britain really be able to survive on her own resources? Will North Sea oil be the answer to her economic problems? Or will the enormous cost of production be a crippling set-back to hopes of future wealth?

In telling the story of North Sea oil, Leith McGrandle examines all the economic, geographical, political and environmental issues involved. He records its history, from the first discoveries of gas and oil off the Dutch coast to the latest finds in the North Sea. He describes clearly the different methods of drilling and bringing the oil ashore, and looks closely at life on an oil rig, as well as in those areas affected by the so-called bonanza.

This book is a valuable contribution to a wide range of studies. It includes a North Sea Timetable, a glossary of oil terms and an index. There are over sixty illustrations and diagrams.

The Story of North Sea Oil

Leith McGrandle

WAYLAND PUBLISHERS

Contents

SBN 85340 433 X

49 Lansdowne Place, Hove, Sussex

Text set in 12/13 pt. Photon Baskerville,
printed by photolithography, and bound in Great Britain at
The Pitman Press, Bath

List of Illustrations

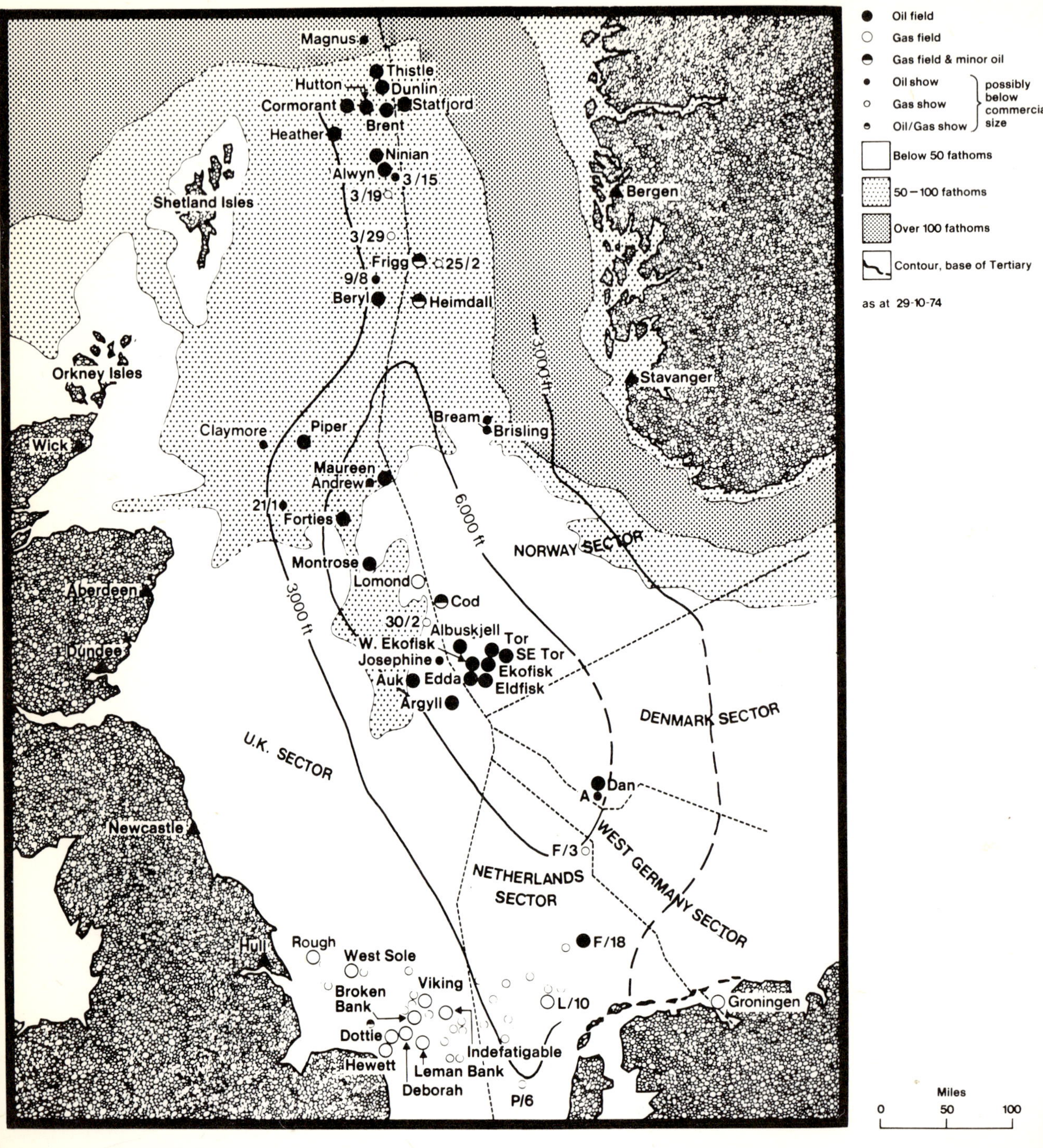
Magnus
Thistle
Hutton
Dunlin
Cormorant
Statfjord
Brent
Heather
Ninian
Alwyn
3/15
3/19
Shetland Isles
3/29
Frigg
25/2
9/8
Beryl
Heimdall
Orkney Isles
3,000 ft
Bergen
Stavanger
Bream
Brisling
Claymore
Piper
Wick
Maureen
Andrew
21/1
Forties
6,000 ft
NORWAY SECTOR
Montrose
Lomond
Aberdeen
Cod
3,000 ft
30/2
Albuskjell
Tor
W. Ekofisk
SE Tor
Josephine
Ekofisk
Dundee
Auk
Edda
Eldfisk
Argyll
DENMARK SECTOR
U.K. SECTOR
Dan
A
Newcastle
F/3
WEST GERMANY SECTOR
NETHERLANDS SECTOR
F/18
Rough
Hull
West Sole
Viking
Broken Bank
L/10
Groningen
Dottie
Indefatigable
Hewett
Leman Bank
Deborah
P/6
Oil field
Gas field
Gas field & minor oil
Oil show
Gas show
Oil/Gas show
possibly below commercial size
Below 50 fathoms
50 – 100 fathoms
Over 100 fathoms
Contour, base of Tertiary
as at 29-10-74
Miles
0
50
100

Introduction

AT FIRST it appears as a tiny speck on the grey-white expanse of sea. Then, as the helicopter loses height and the blur of the sea below forms into long ripples of waves, the speck starts to grow and take shape. It looks as though someone had started to build an Eiffel Tower out of a giant Meccano set, but ran out of pieces half way through. Ungainly struts of grey metal stand out of the sea. As the helicopter comes into land on the pad, a large, yellow-painted circle flanked by safety nets, a few tiny figures can be seen scuttling about among a mass of pipes. Even at close quarters, the modern oil rig, far from land in the middle of the North Sea, looks unimpressive and vulnerable. Yet this is where the story of North Sea oil begins. On this drilling rig, and on the forty or more dotted around the thousands of square miles of the North Sea, live the men who probe the sea bed for tell-tale signs of oil and gas.

Already the oil discoveries in the North Sea have surpassed many an oilman's wildest hopes. Until 1975 Britain produced hardly any oil. All she needed – for running cars or lorries, heating homes and factories, and converting into electricity – came from the Persian Gulf, Nigeria, Venezuela or other distant countries. In 1974 she used over 100 million tons of oil, and the bill came to around £3,000 million.

Today Britain is producing oil of her own from the North Sea. It is still a tiny trickle – two million tons so far, or less than one fiftieth of her needs for 1975. But by

Opposite page A map of the North Sea showing the oil and gas fields discovered by the end of 1974.

Above Israeli soldiers march into battle during the Middle East War of October 1973. This war had a severe effect on the supply of oil to Europe and America.

1980 she could be producing all the oil she is likely to want in a year, and even have some left over to sell abroad. Beneath the sea bed is enough oil to keep her going for perhaps 30 to 40 years.

The North Sea oil boom is an unexpected blessing. Britain has never been rich in minerals. Coal and iron ore have so far been the only two she has had in quantity, and on these were built the nineteenth century Industrial Revolution. Now Britain has discovered a third mineral – oil – at a time when oil has taken on a crucial importance in the world. Oil is the lifeblood of modern industry. We use it as petrol for cars, and from it comes diesel fuel for our buses, lorries and trains. It fuels many power stations producing electricity. It can be turned into plastics or artificial fibres, or many kinds of chemical. And, until recently, it was cheap. It is little wonder that oil in some form or other is used in almost every part of our life. Without it, industry would seize up. The disruption caused when oil supplies were cut off immediately after the Middle East War in October 1973 proved that. Now people are prepared to pay almost any price for it.

Oil is twentieth century man's new gold. For its owners it can bring riches beyond the dreams of earlier generations. And it can bring power, too; oil under the sea or sands can be as mighty as armies in the world's balance of power.

Of course all countries can counter this power, simply by using less oil. This has been shown by the economy measures taken since the five-fold increase in the price of oil made after the last Middle East War. But oil still remains overwhelmingly important and will continue to do so. There are other sources of energy, but it takes a long time to develop them. Ten years can pass between the decision to start a new coal field in Britain and the day when it reaches full production. It can take even longer to put a nuclear power station into operation. Even an oil field in the North Sea can rarely come onto full flow in much less than eight years.

Some alternative sources of energy have a few of the

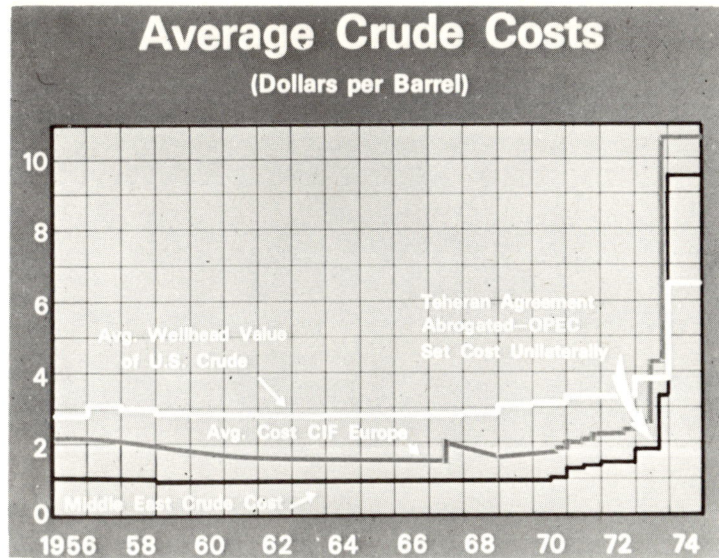

Left The price of crude oil (in dollars per barrel) from 1956 to 1974. This graph clearly shows the drastic increase in the price of crude oil after the Middle East War in October 1973.

advantages of oil. Both natural gas and nuclear power can produce energy in the form of electricity, and could replace oil. But neither is a substitute for oil in making chemicals or plastics. Coal can be used as a source of both power and chemicals, but it has been more expensive than oil to break down into other materials. In 1973 almost half of Britain's energy needs were met by oil. Around 40 per cent came from coal, some 10 per cent from natural gas and the rest, a relatively small amount, from nuclear and hydro-electric power. If our energy demands continue to increase at the same rate as over the past few years, even given a gradual switch to other energy sources and a reduction in the rate at which we consume oil, we shall still be using more oil in 1980 than we do today.

At the moment oil provides 60 per cent of Western Europe's energy. The total demand for crude oil in the area currently amounts to some 15½ million barrels of oil a day. Even after the rise in oil prices since the last Middle East war, estimates put the demand for oil in Europe in the early 1980s at around 20 million barrels a day. So Europe, and the rest of the world, will continue to need a great deal of oil, and most of it will still come

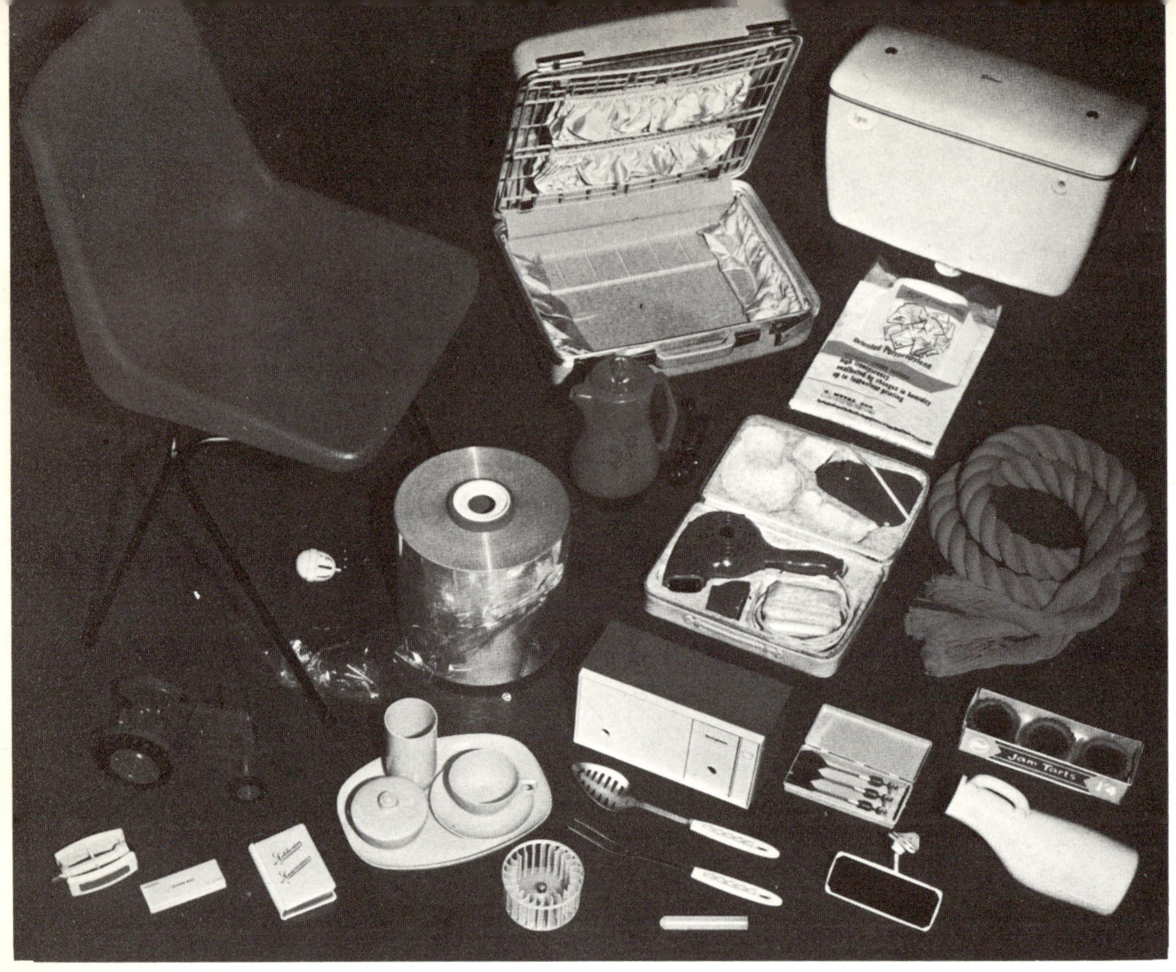

Above It is not only factories, ships and cars that depend on oil; the plastics in this picture are all made from oil-based chemicals.

from the Middle East. For the hard fact remains that three-quarters of the world's known oil reserves totalling nearly 400 billion barrels) are in the Middle East, and well over half are in one country – Saudi Arabia. Compared with those massive amounts of oil Britain's North Sea finds – even at the highest estimates – will form only about 3 per cent of the world's oil reserves by 1980. They will also cost much more to develop. The investment needed to drill a well in the Middle East is around £100 per daily output of one barrel. In the North Sea it can be more than £1,400.

Even so, North Sea oil remains of crucial importance to Britain. Current estimates put the recoverable

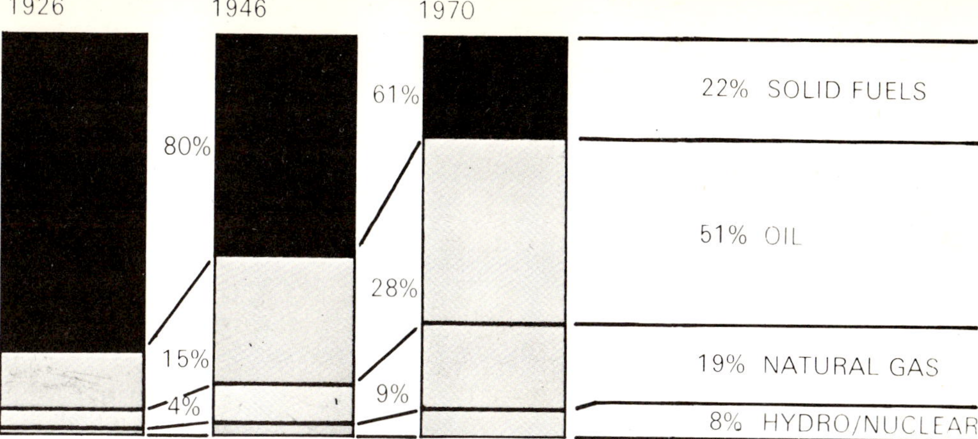

Above The world's energy sources fall into four main groups. This diagram shows the increasing importance of oil over the last fifty years.

reserves of oil at over 18,000 million barrels (over 3,000 million tons). By 1980, provided technical and – more important – political factors are favourable, then Britain could be producing between 120 million and 150 million tons a year. In the words of a Department of Energy report in May 1974, "there is now a very good chance that in 1980 we can produce oil equivalent to our demand."

Moreover, North Sea oil is a special type of oil, a light oil with a low sulphur content. Crude oil, in fact, is made up of many different liquids of varying weight, known as fractions, and each having a particular use. The heaviest fractions are fuel oil, for ships, industry and central heating, and bitumen, used for surfacing roads and waterproofing. The lightest fractions are refinery gas, used as fuel in oil refineries, and petrol. North Sea oil is made up of these lighter fractions. Being a rare kind of oil, it is in great demand all over the world. But Britain needs other types of oil, too, such as the heavy oils which are found in the Middle East. But selling our own special kind of oil will help to pay for the oil we need to import.

Given the importance of oil to an advanced economy like Britain's, and considering its rising cost, North Sea oil is obviously a major economic windfall. Already, in a few short years, it has made its presence felt. Rigs exploring for oil roam the North Sea. Giant platforms for sucking the oil out of the sea bed are now in position, or are being built on the coasts. Industries on which the

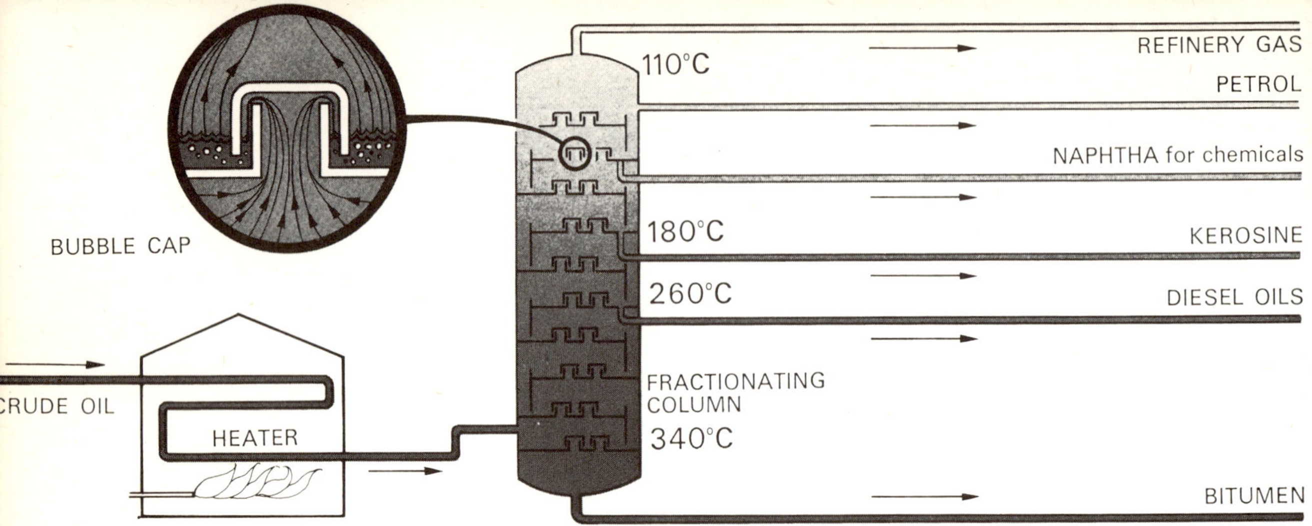

Above In a refinery crude oil is split up into fractions by the process shown in this diagram. Each fraction is of a different weight, and has a different use.

Opposite page North Sea oil is a light oil, made up of light fractions. This diagram shows the products made from these light fractions.

oil industry depends – such as rig-building – have sprung up. Towns such as Aberdeen and Montrose in Scotland have been touched with the Klondike spirit. New jobs have been created and fresh fortunes made.

Yet with the oil have come problems too. They are not solely technical – there are, for example, the dangers facing the oilmen in some of the worst seas in the world, the disappointments when an oil-well proves to be "dry", and the difficulty of raising the vast sums of money needed to finance the projects. There are human conflicts, too. In Scotland, in particular, the oil boom has brought a clash between the old ways and the new. It has changed towns, created additional pressure for housing and threatened the character of some areas. The need to build industrial sites has sometimes been met with opposition from the local people. Oil has created political headaches, too. It has fuelled support for the Scottish Nationalist Party, who believe that the oil is Scotland's oil, and that only Scotland should benefit from it. In England politicians have grappled with the question of where the riches expected from the oil should go and in what amounts. But many of these problems only arise once the oil reaches land and after it flows in quantity. First, you must find your oil. . . .

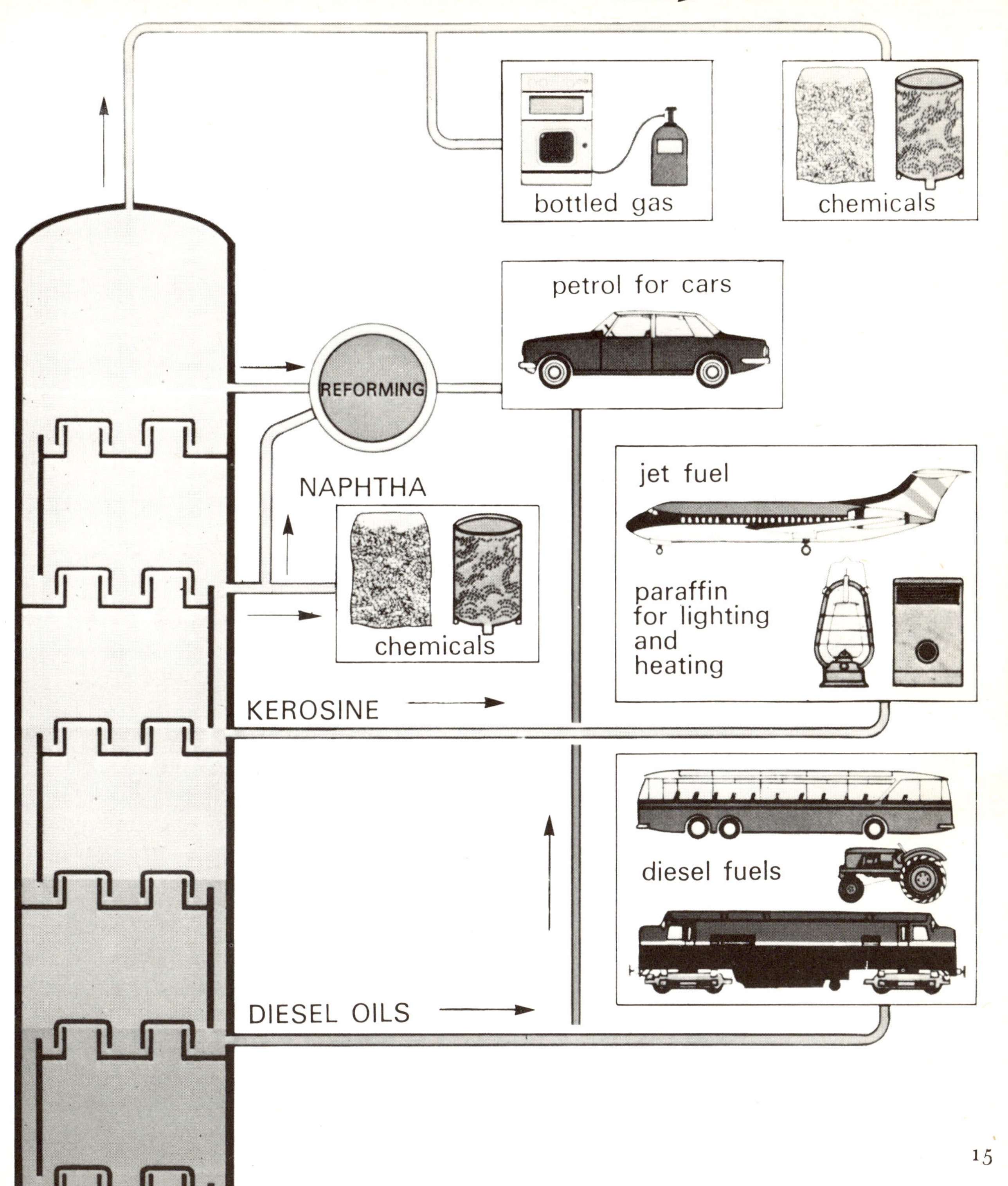
bottled gas
chemicals
petrol for cars
REFORMING
NAPHTHA
chemicals
jet fuel
paraffin
for lighting
and
heating
KEROSINE
diesel fuels
DIESEL OILS

1. Early Days in the North Sea

FOR THE WORKMEN on the oil rig perched in shallow waters off the tiny Dutch village of Schlochtern, 15th August, 1959 was just an ordinary day. They had been drilling in the area for some weeks. This was yet another routine borehole. But that Saturday, at nearly 3,500 feet underground, they struck gas. Neither they nor the company they worked for, Nederlandse Aardol Mij, knew it at the time, but they had stumbled on an enormous reservoir of natural gas – the second largest in the world. They had triggered off the first step in the search for North Sea oil. Gas and oil are nearly always found together. If, the geologists argued, there was gas under the North Sea, there would also be oil.

But in those early days the main interest was in gas. The Dutch discovery gave British geologists the idea of exploring for natural gas in the southern half of the North Sea. They were encouraged by the fact that small pockets of gas had already been found near Whitby in Yorkshire. So the oil companies set to work. Over the next five years they spent around £10 million on surveys, and built up a picture of the geological structure of the sea bed.

The sailor's eye is trained to observe the movement of the tides, the waves and the stars; the diver's to see the plant life, the fish and the darkness of the deep. But the geologist probes beneath all this. To his eye the sea bed looks like one of those modern paintings splashed with blobs of bright colour, each blob representing hundreds

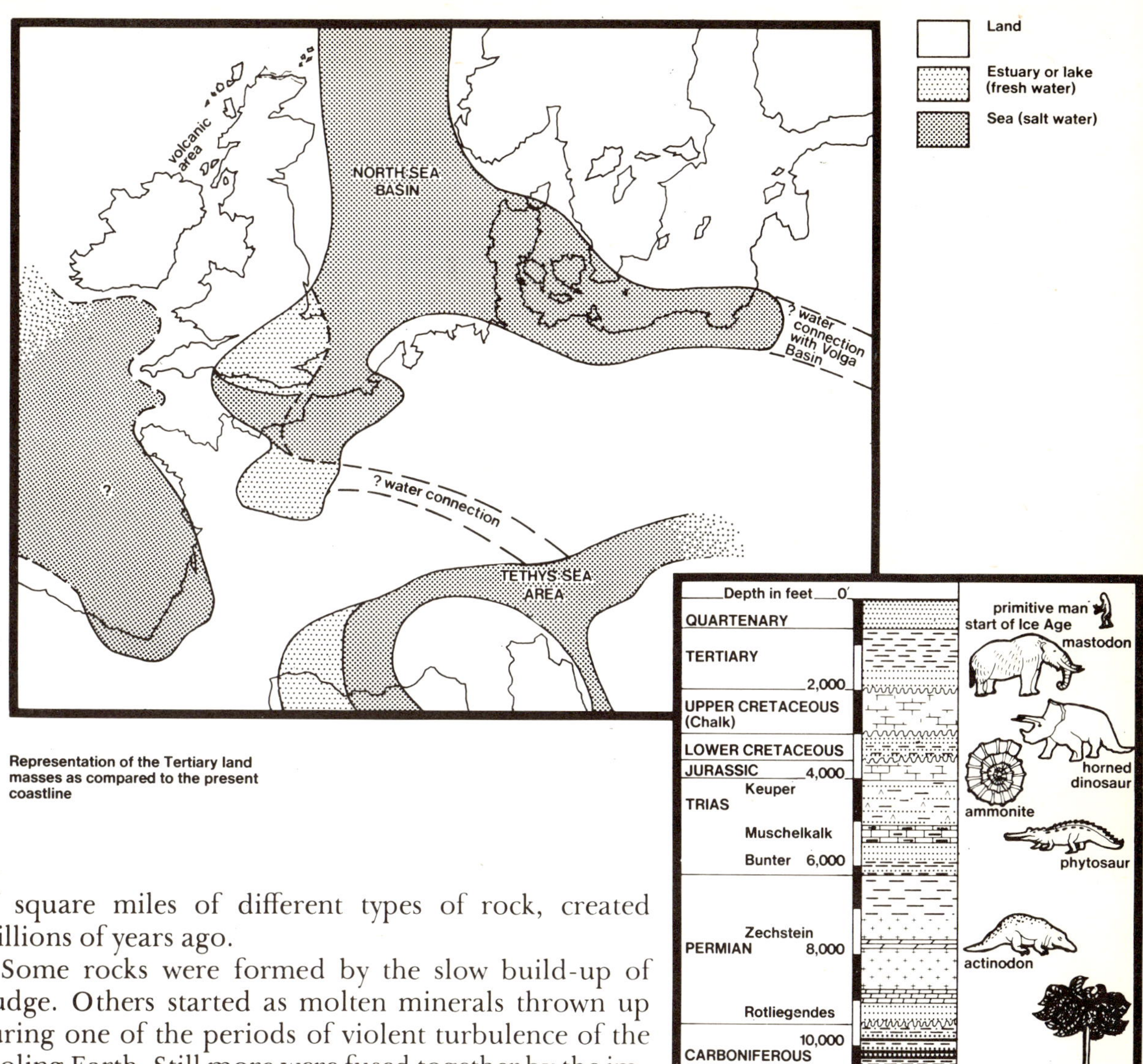

Representation of the Tertiary land masses as compared to the present coastline

of square miles of different types of rock, created millions of years ago.

Some rocks were formed by the slow build-up of sludge. Others started as molten minerals thrown up during one of the periods of violent turbulence of the cooling Earth. Still more were fused together by the immense pressure created as the continents shifted and settled in prehistoric days.

Some of the layers of rock below the North Sea are more than 600 million years old, others a mere 25 million. Within these layers of rock lie rich reserves of oil and gas. At least three enormous basins rich in oil and gas lie under the British part of the North Sea. They

Above The land masses in the Tertiary age (between 10 and 60 million years ago). The North Sea evolved from a large area of water existing then.

begin with the large gas-bearing rock structure off the East Anglian coast, which stretches in a curve up towards Lincolnshire and the East Riding of Yorkshire. Secondly there is the large complex of oil in the midst of the North Sea, about 200 miles off the Aberdeen coast. This includes the famous Ekofisk field in Norwegian waters and the Forties oil fields. Finally there is the newer area of exploration, half way between the Shetland Islands and Norway, which contains the Brent oil field and Frigg gas fields.

Some of the oil found off the North Sea is trapped in rocks formed as much as 270 million years ago, but most of it was formed between 70 million and 200 million years ago. (The first humans are believed to have developed on the planet about three million years ago, and our earliest civilizations go back only about 7,000 years.)

The oil in the North Sea began to be formed millions of years ago, when countless tiny animals and plants in the sea died and sank to the sea-bed. Gradually mud was washed down to the sea by the rivers and became mixed with the decaying matter. Over millions of years, layer upon layer of mud built up, until the lower layers were formed into rocks by the enormous pressure of weight above them. During this process, oil and gas were formed from the rotting plants and animals. (It is because this process took so many millions of years that, once we have pumped out the existing supplies of oil, they can never be formed again.)

As further pressure built up, the oil and gas were squeezed upwards through porous rock (rock containing tiny spaces, allowing liquids to filter through), until they reached non-porous rock (solid rock). They then remained trapped inside the porous rock forming pools of oil. (The term "pool" is misleading; the oil does not form a large lake, as it suggests, but is contained within the rock itself, like water in a sponge.) The gas mixed with the oil rose to the surface of the pool, and it was into such a layer of gas – a gas cap – that the Dutch oilmen drilled on that summer day in 1959.

Opposite page: top Oil collected in porous sandstone, in a fold of rock formed by constant pressure over millions of years. *Inset* A detail of the oil-bearing sandstone.
Bottom Drilling into a pool of oil, gas and water.

2. First Came the Gas

ENCOURAGED BY their geological reports, the big oil companies started to move in to explore for natural gas. Surveys could give them a clue to where oil and gas might be. But until they actually started drilling down from a rig into the sea bed, they could not be sure.

By Christmas 1964 the first drilling rig, known as *Mr. Cap*, operated by Amoseas (American Overseas Petroleum Ltd.), was "spudding in" – starting on its first drillings – on the Dogger Bank. The beginning was disappointing. By the summer of the following year, three wells had been drilled, but all were found to be dry. It was a grim reminder that the search for oil and gas is long and frustrating. But in December 1965 British Petroleum's rig, *Sea Gem*, discovered a large gas field off the estuary of the River Humber. The official name of the field, taken from the area licensed from the Government, was block 48/6, but it became known as West Sole.

A few months later three more major gas fields were discovered off the Norfolk coast. Together with West Sole, these three – the Leman Bank, Indefatigable and Hewett – were estimated to hold enough gas to satisfy Britain's needs until the 1990s. A little over two years after the first discoveries in the British sector of the North Sea, the first gas was piped ashore from the West Sole field to Easington in Yorkshire.

By 1969 178 wells had been drilled and five major gas fields discovered. The Gas Council began to switch from town gas to natural North Sea gas. A massive complex of

Above Opening the pipes to the full flow of gas from the North Sea.

Opposite page BP's North Sea Gas Terminal at Easington, in Yorkshire. The first gas to be piped ashore from the North Sea was stored here.

Above A gas production platform in the Viking field.

pipelines and gas terminals was constructed. The conversion of everything from factory heating systems to the domestic cooker began. By 1975 almost 86 per cent of gas users in the country were using North Sea gas. Now some 3,500 million cubic feet a day is being piped out through the 600 miles of gas line below the sea. And there is more to come. By 1980 the Gas Council expects all gas consumers to be using natural gas. On present estimates there is still enough gas to last for at least 20 to 30 years, taking into account new discoveries in the North Sea.

3. Then Came the Oil

EVEN AS they searched for gas the oil companies were on the hunt for oil. Gas was valuable, but oil was the really big prize. Gas prices were often restricted by government. The British Government, for example, had set a price of around 3½p a therm to the producers of North Sea gas. (The therm is the standard unit of heat value in gas supply.) But the oil prices were set by world markets and negotiation. So there was a greater incentive to search for oil.

Moreover, the world seemed to be running out of oil – at least in the long run. By the late 1960s, two-thirds of the world's energy consumption was of oil or natural gas. There was still plenty of oil, especially in the Middle East. But man's need for it was growing faster than his ability to discover new oil fields. The United States, for example, at one time had more than enough oil to satisfy its population's needs, but was reaching the stage where it would have to import oil. Even more disturbing, the rate of discovery of new wells was tailing off. Not only was exploration becoming more difficult, but it was also getting more expensive. The oil companies were being forced faster and farther into harsher climates.

Geologists started to brave the harsh Arctic weather of Alaska where, in winter, the temperature can fall to minus 70°F, and where it is dark for almost six months of the year. But perseverance brought success. In 1969 large oil fields were discovered around Prudhoe Bay in

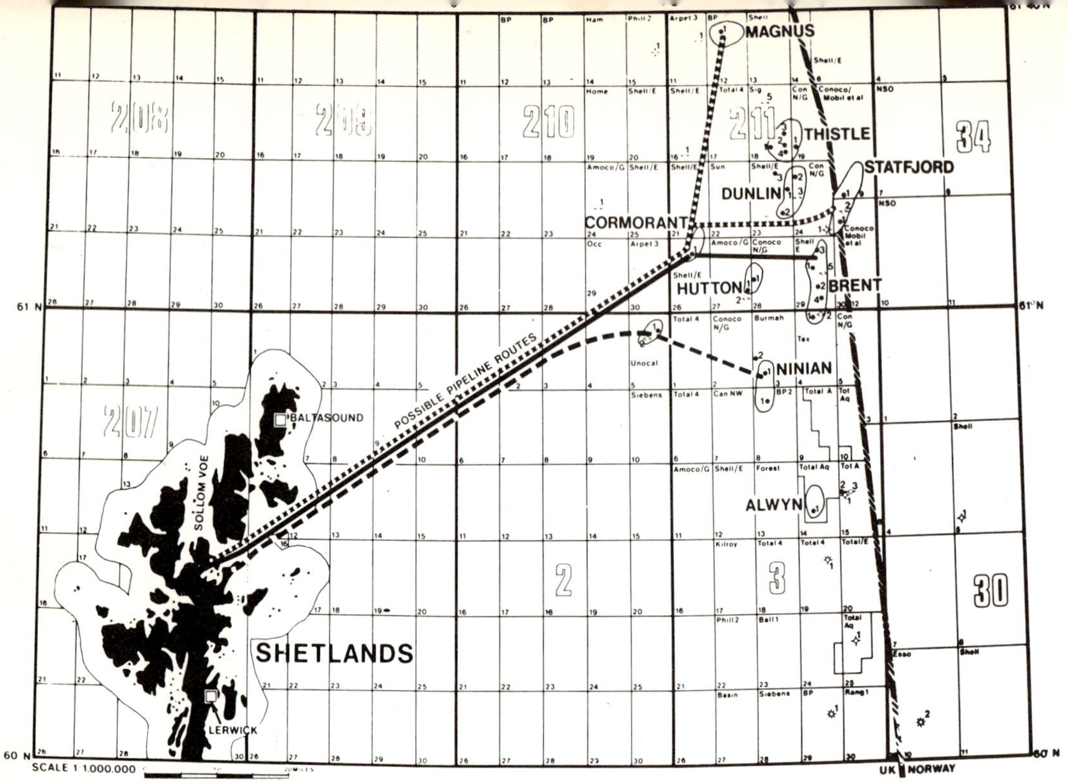

Above A map of the North Sea off the Shetland Islands, showing the oil fields discovered since 1970. This map also shows how the North Sea is divided into blocks, each one allocated to one oil company for exploration and drilling.

Alaska. Beneath the permanent ice lay oil reserves large enough to supply the United States with oil for over twenty years. And then there was the North Sea. Its waters were deeper and wilder than any in which oil had been searched for before. But the world's need for more oil, and the wealth it would bring if it was found, spurred the oil companies and governments on.

In December 1969 a consortium led by an American oil group (Amoco) and a British nationalized industry (the Gas Council) announced that it had struck what appeared to be an oil field 150 miles off the Aberdeen coast. The field had "commercial possibilities": that is, there was enough oil to make it worth spending the vast sums of money needed to pipe it out of the ground and across possibly hundreds of miles of sea bed. The field was later christened "Montrose."

A few months later, in June 1970, the Norwegians uncovered a "giant oil find" in their sector of the North Sea. This Ekofisk discovery encouraged the British explorers, for it suggested that they could be on the right trail. They were; in mid-October 1970 British Petroleum struck the United Kingdom's first major oil field – the Forties Field – 120 miles off Aberdeen. The Forties Field was estimated to contain 1,500 million barrels of oil which could be brought to the surface (this estimate was later raised to 1,760 million). It was big enough to produce every year a fifth of Britain's demands at that time.

The search in Scottish waters speeded up. In mid-1971 the Royal Dutch Shell Group, together with Esso Petroleum, revealed that they had struck oil in block 211/29, about 120 miles off the Shetland Islands. This field, called Brent, was further north than any previous North Sea find, and has now proved to be part of one of the biggest oil fields in the world. From then on the discoveries flowed quick and fast. In 1972 the Thistle field was discovered. The following year brought three further fields, Piper, Dunlin and Hutton, with recoverable reserves of 3,000 million barrels, and the ability to produce around 600,000 barrels a day, a quarter of the country's present needs. By January 1974, after some disappointing dry wells, there were encouraging reports of a massive oil field – Ninian – ninety-five miles east of Shetland. Early reports suggested that Ninian was at least as big as the Forties Field, and might be much bigger.

In five short years, the results had been remarkable. Estimates of the recoverable oil in the North Sea now range between 18,000 million barrels and 50,000 million barrels. This still seems small compared with the total reserves of the Middle East, which are nearer 400,000 million barrels. Yet this 3 per cent of the world's oil supplies so far discovered is of great strategic importance, especially as Western Europe cuts down its dependence on Middle East oil. The exploration goes on. In October 1974, British Petroleum revealed a new oil field (Andrew) off the Shetlands. By the end of 1974 there was a record number of oil rigs drilling in the North Sea – twenty-five of them in the British areas alone. New discoveries were made in 1975.

4. Exploring the Deep

THE NORTH SEA is the last place any oilman would want to search for oil. During the winter months, which can stretch from early October into April, waves some 75 feet high smash against the legs of the oil platforms. Gale-force winds can exceed 100 miles an hour, whipping the sea into a wild frenzy. In the winter of 1973–74, for example, there were three particularly savage gales, two in November and one in January, which produced 60 to 70 foot waves and 65 mile-an-hour winds.

For as much as a tenth of the year the wet North Sea fog cuts visibility down to less than half a mile. All the exploration, therefore, has to be crammed into the spring and summer months before the "weather window" closes. Even then, gales can spring up suddenly, fog may lurk for days and the exploration rigs can be cut off from outside help.

The quick-changing moods of North Sea weather are a forecaster's nightmare, yet accurate forecasting is vital. With the costs of running an offshore rig at over £20,000 a day it is an expensive business to stop work if a weather forecast is proved wrong. But shutting the rig down too late is even worse, as expensive machinery could be damaged by the high seas. So the forecasters have found themselves with a great responsibility thrust upon them.

Three weather ships form a key to this forecasting. One operates in the North East Shetlands, the second in the West Shetlands area and the third in the Irish Sea. Each sends back reports every three hours covering wind

Opposite page: top *Glomar V,* a drilling ship used by Mobil Oil for exploratory drilling in the North Sea. *Bottom* A geologist examines rock samples from the sea bed for traces of hydrocarbons; these indicate the presence of gas and oil.

speeds, heights of waves, barometric pressures, and general weather conditions. Also, the oil rigs themselves now send back weather reports to the London Weather Centre. The result is that weather conditions in the North Sea are plotted in greater detail and with more accuracy than ever before.

But if the North Sea presents headaches for the weather forecasters they are nothing compared with the task facing the scientists. Oil has never been sucked out of waters as wild or as deep before. Some of the main fields lie below 600 feet of water. Until the oil companies started exploring in the North Sea, the deepest water they had drilled in was 250 feet. The sea bed itself is shifted around by powerful tidal currents that can carry enormous quantities of sand from one area to another. The problems of trying to keep a rig stable on such a sea bed are formidable.

Exploring and developing the North Sea has forced the scientists, technologists and engineers to develop a new range of technology. In some areas the scientists are wrestling with problems as difficult and unpredictable as those found in the United States space programme. They are working on the frontiers of science.

One area of new research has centred on the great steel and concrete production platforms. These are often 700 or more feet high, and are driven to as much as 250 feet below the sea bed; they can rest in as much as 500 or 600 feet of water. Extensive work has had to be carried out by soil mechanic experts using new techniques to gauge the depths, structure and types of soil in the subsea bed on which the platforms rest.

The oil rigs and platforms themselves must be designed to withstand the terrible weather conditions, and this has posed new problems for stress engineers. For example, a 60-foot wave hits a steel production platform with a force of 140 lbs. per square foot. A 70-foot wave smashes against it with a force of 185 lbs. per square foot, and a 100-foot wave (never yet experienced against a platform but it has to be allowed for) would crash against it with a force of 300 lbs. per square foot.

Opposite page A diving bell being lowered for deep diving operations at night.

New steels and stronger types of welding have had to be developed to cope with these stresses. In order to drive in the huge 250-foot piles of an oil platform, engineers have built the world's largest pile hammer. The offshore pipelines have been specially made. For example, special concrete mixes for the coatings surrounding the pipes have been produced which include iron ore and millions of small steel reinforcing fibres.

The manufacture of heavy concrete production platforms has meant building the largest derrick barges in the world; these can lift 2,000 tons in a fixed position, or 1,500 tons when revolving.

Diving operations, which are hazardous but essential at many stages in North Sea operations, have also led to new technologies. New diving bells and small decompression chambers have been designed to enable divers to work at the unprecedented depth of 600 feet. Midget submarines, rather than diving bells fixed to a mother ship, are the next stage in deep water technology, and beyond this lies the possibility of subsea stations (see Chapter 16).

The costs match the risks and the new techniques. Over £1,500 million has already been spent in the British sector of the North Sea; this sum is expected to grow to at least £10,000 million by 1980. The cost of building one floating drilling rig is at least £20 million. Together with services needed to maintain it, such a rig costs around £30,000 a day to run.

A day lost by bad weather is wasted money, and time gone for ever. Each exploratory well can cost more than £1 million to drill (one in the winter of 1972–73 cost £2 million) and costs are rising all the time. Even at the end of all this, the drilling could turn out to be a waste of time and money if the well turns out to be "dry." In December 1974, for example, British Petroleum announced that two wells about 130 miles west-south-west of the Shetlands had been abandoned. The total loss to the company was £5.6 million.

During 1973 only one well in every eight drilled struck oil, and this is a high success rate in the oil business. In

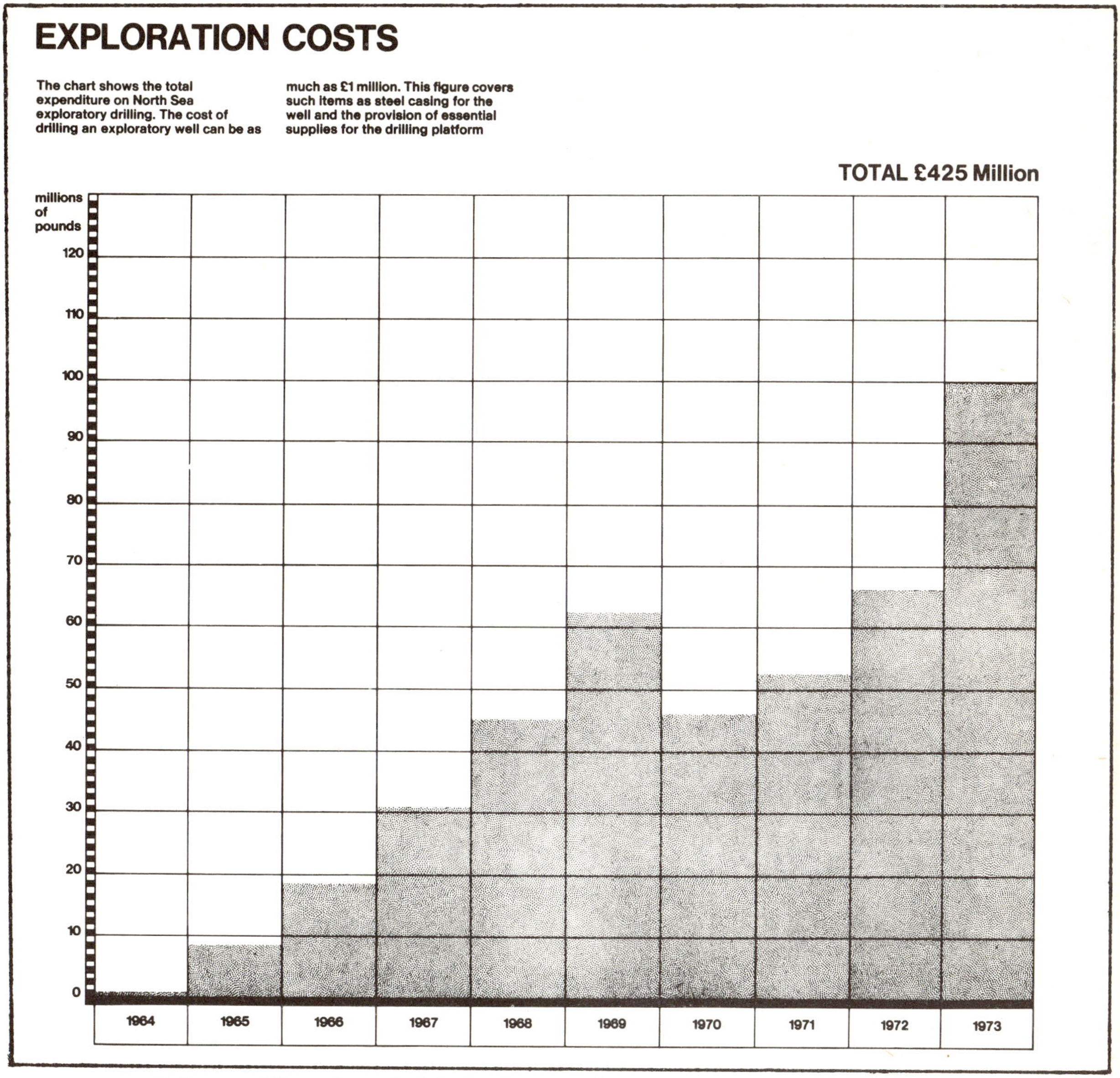
EXPLORATION COSTS
The chart shows the total expenditure on North Sea exploratory drilling. The cost of drilling an exploratory well can be as much as £1 million. This figure covers such items as steel casing for the well and the provision of essential supplies for the drilling platform
TOTAL £425 Million
millions of pounds
120
110
100
90
80
70
60
50
40
30
20
10
0
1964
1965
1966
1967
1968
1969
1970
1971
1972
1973

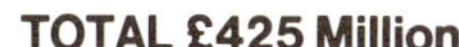

many parts of the world, and probably in the North Sea as the oilmen move into less promising areas, the failure rate will be much higher.

Only our need for oil (and the high price we are prepared to pay for it) has forced the explorers to spend such astronomical sums in such inhospitable waters.

Above The rising cost of exploration in the North Sea.

5. The Types of Rig

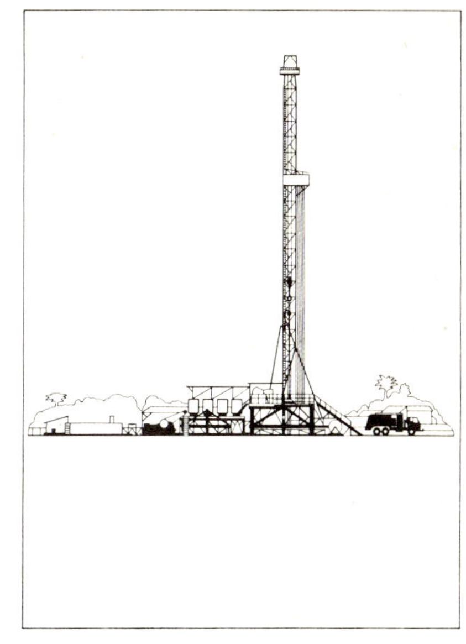

Above A land drilling rig.

OFFSHORE OIL EXPLORATION has been known in California since the turn of the century, but it started on a substantial scale in the mid-1930s with a few pioneering efforts in the marshy waters of the Mississippi River Delta. The drilling rig, the traditional symbol of the oil industry, was a crudely-built wooden affair. Some minor adaptations were made to allow for the move from dry land to water. Used initially in 20 feet of water, the rigs rested on the river bed.

As the search moved into deeper water, around the coast of the Gulf of Mexico and Venezuela, the fixed platform rigs, as they were called, became more complicated. In depths of 50 to 60 feet they became useless, and a new type of rig – the "jack-up" – was developed.

The jack-up rig is a platform with enormous legs which can be telescoped together. While it is being floated from one drilling spot to another its legs are retracted. Once the rig arrives at the new site the legs are dropped down to the bottom of the sea bed. The drilling platform is then jacked up to the surface rather like a car being jacked up when a tyre needs changing. Jack-up rigs, which cost £6 million to build, can operate in up to 300 feet of water.

At the heart of any oil rig is the drill which is pushed down into the sea bed to test whether or not the well is dry. Whatever the size or type of rig an oil well, whether at sea or on land, is basically the same. It is, quite simply, a hole. It can be anything between 2,000 and 30,000 feet

Below A jack-up rig.

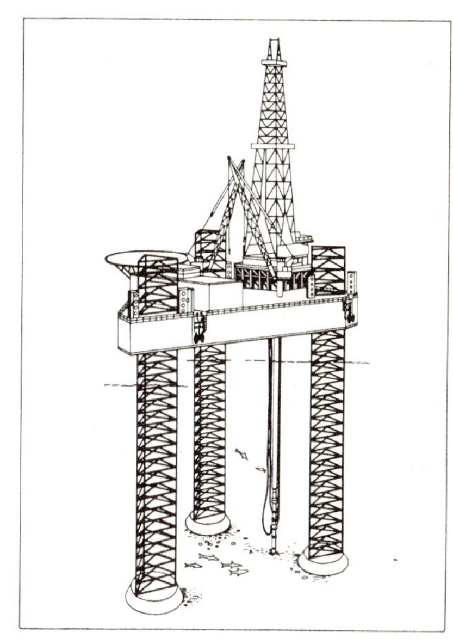

Opposite page Towing a jack-up rig to its new position with its legs jacked up.

deep and 7 to 30 inches or more wide. The sides are supported by a series of steel pipes cemented in and known as casings. The drilling of the well is done by using a long tube of steel (the drilling string) tipped by a rotating bit with steel or diamond teeth. As the drilling goes deeper the drilling tube or string is lengthened by adding another section of tubing, rather like adding another length of support to a tent pole.

During drilling a heavy fluid, known in the oil world as mud, is pumped down the hollow centre of the drilling pipe or string. This "mud" is a sludgy mixture of clay, water and certain chemicals. Driven down the

Below: left Lengths of drilling pipe, or string, stored ready for use.
Right The drilling bit being drawn up from the sea; the teeth of the bit can be clearly seen.

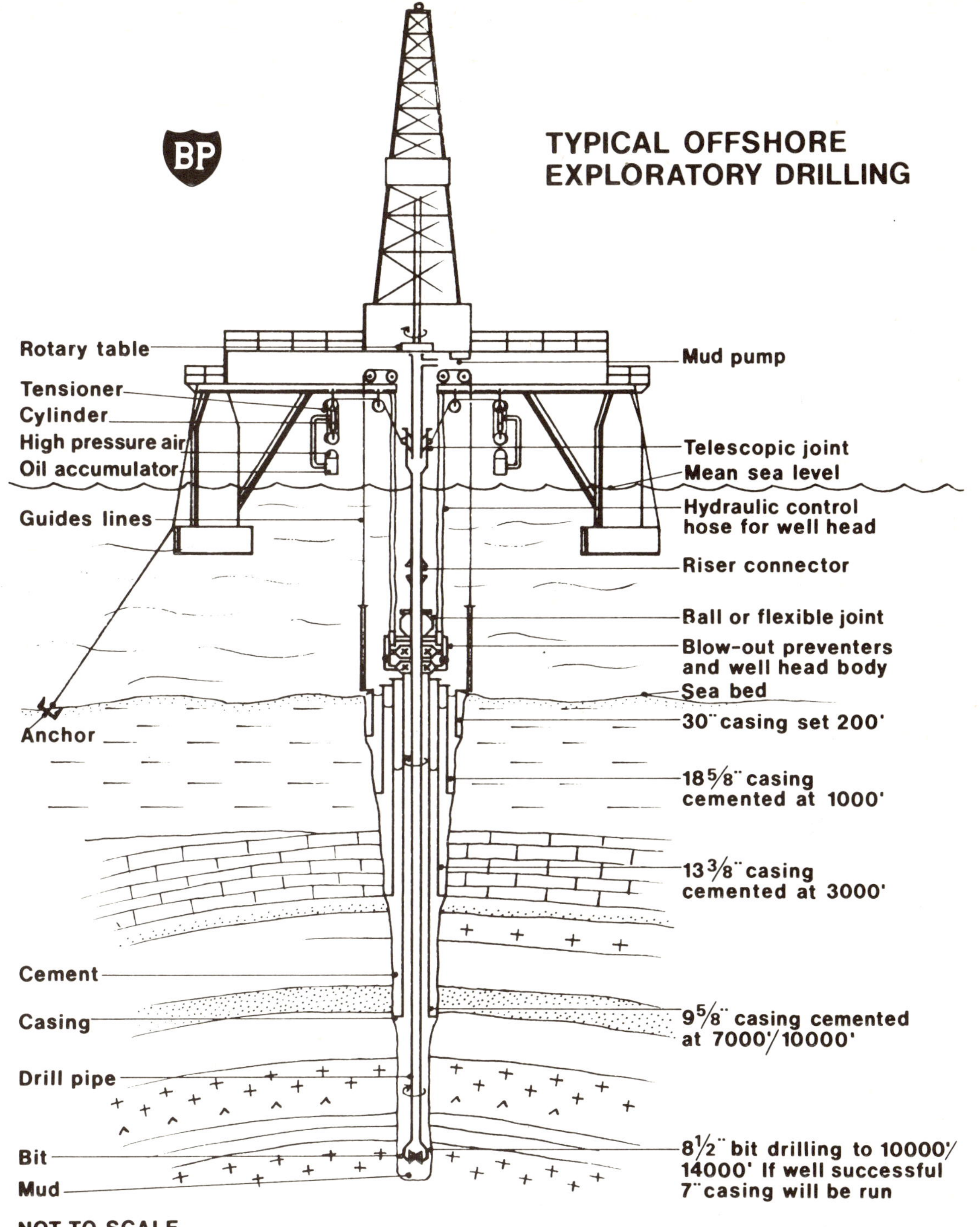
BP
TYPICAL OFFSHORE
EXPLORATORY DRILLING
Rotary table
Tensioner
Cylinder
High pressure air
Oil accumulator
Guides lines
Anchor
Cement
Casing
Drill pipe
Bit
Mud
Mud pump
Telescopic joint
Mean sea level
Hydraulic control
hose for well head
Riser connector
Ball or flexible joint
Blow-out preventers
and well head body
Sea bed
30" casing set 200'
18 5/8" casing
cemented at 1000'
13 3/8" casing
cemented at 3000'
9 5/8" casing cemented
at 7000'/10000'
8 1/2" bit drilling to 10000'/
14000' If well successful
7" casing will be run
NOT TO SCALE

Above On the drilling platform: the drilling string rotating during drilling.

hollow drill pipe, it comes out through holes in the drilling bit and is pumped back up to the surface through the gap between the drilling string and the steel casings. The mud carries with it fragments of rock cut away by the drilling bit. These rock fragments help the geologists to build up a detailed picture of the geology of the layers of rock through which the bit passes. The geologists know they are really onto something when the fragments of rock contain traces of hydrocarbons.

When the drill bites through into an oil pool the pressure of gas and oil is sometimes so great that it jets up to the surface. This is the origin of the "gusher." In the old days, before the oilmen knew how to bring the gusher quickly under control, it could pour out like a fountain for hours, even days. Now, the "gusher" is usually dampened rapidly. In many cases there is too little pressure to push the oil up to the surface and it has to be pumped up.

The platform of the oil rig is fairly simple. At the top of the hole is a rotary table which turns the drilling pipe. There is equipment for pumping and collecting the mud. A derrick is needed to help raise and lower the drill pipe. Then there are storage tanks, mud separators and the crew's living quarters, which are usually tiny. A helicopter pad, often saucer-shaped, has become an essential part of rigs drilling far from the coast.

The first hints of oil show up on the "mud logger," which indicates the amount of hydrocarbon (a sign of oil) in the mud. Further tests are made to see how much oil there is in the rock. One successful well is not enough to make an oil company decide to go ahead. At least four, sometimes many more, must be drilled to see whether the field is worth exploiting. Only then will the company decide to develop the field with a massive fixed platform designed to pump the oil up from the sea bed.

The jack-up rig has been a great success in many waters, but the fierce North Sea proved too much for it. Even where the sea was not too deep for it, the shifting sands on the sea bed made the jack-up rig impractical. In the early days of drilling the oil companies used

A semi-submersible rig.

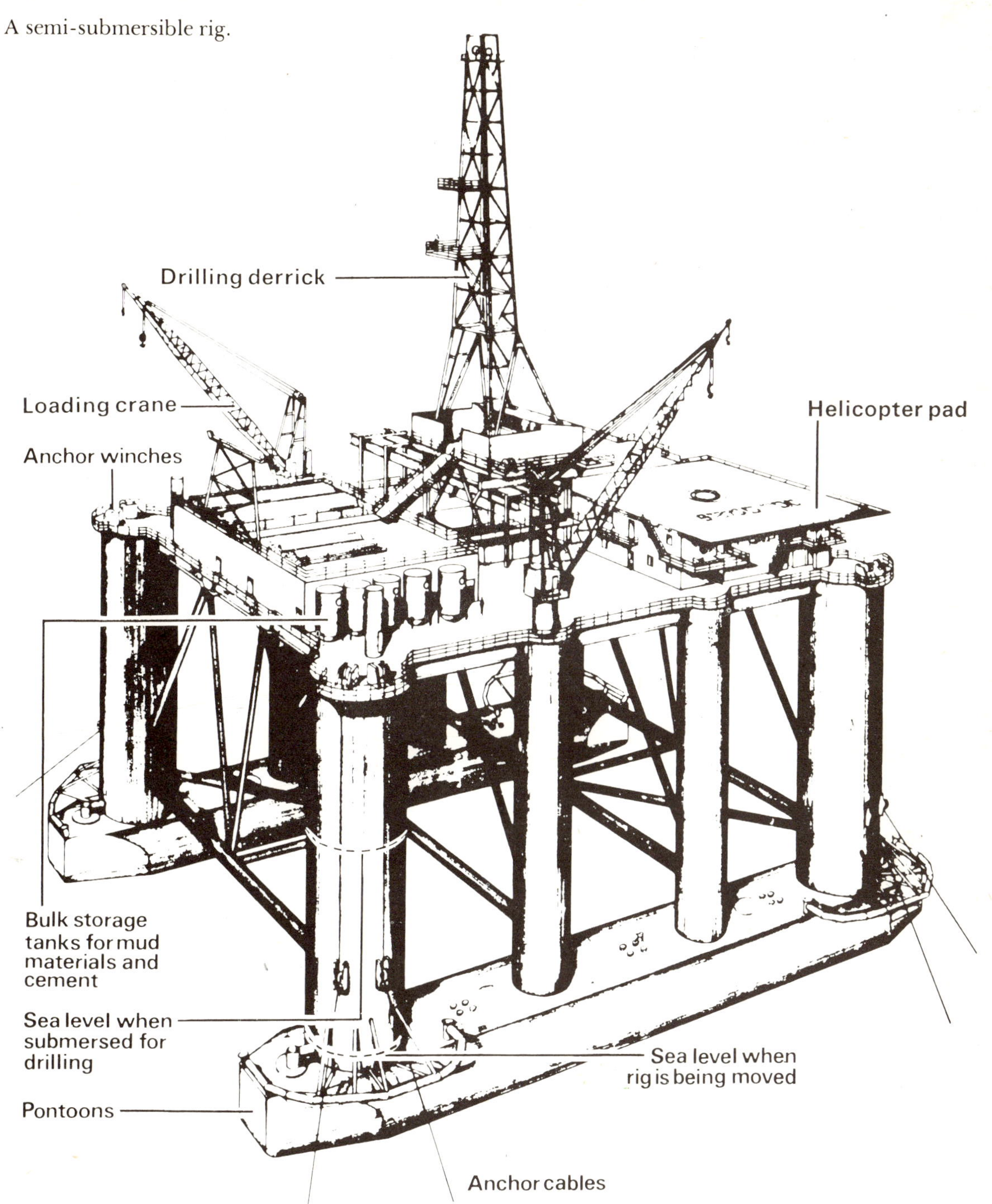

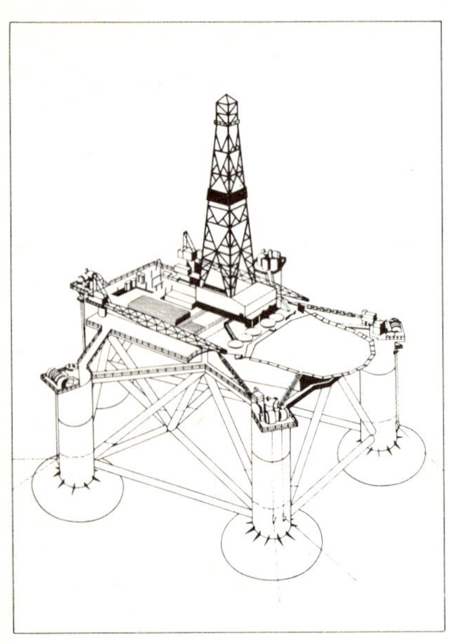

Above A semi-submersible rig.

drilling ships in which the drill is put down to the bottom from a ship specially designed for the purpose. This avoids having to rely on steel supports resting on the sea bed. But it was found that, especially in rough northern waters, the time during which a drill ship could be used was very limited.

The oil companies have found it worthwhile to spend the extra time and money on developing semi-submersible drilling rigs. With these, the drilling platform and drill camps complete with housing, mess hall and workshops, rest on huge steel legs attached to submerged tanks. These give buoyancy and provide stability even during the wildest of gales. The great drawback is their expense – £20 million each – but they have proved essential in the North Sea. The most advanced semi-submersibles now being built will be able to drill both exploratory and production oil wells in waters more than 2,000 feet deep.

Supplying the oil rigs in the North Sea has created a new industry. Rigs, often operating a couple of hundred miles offshore, have to be kept well equipped with food and fuel, "mud," water, chemicals, casing pipes and other equipment. These have to be carried from the shore by boats able to cope with rough weather, and then transferred to the rig platform. The platform can be as much as 100 feet from the surface of the sea and a boat might be rising and falling in the waves by as much as 20 or 30 feet.

Strong, robust boats (which can stand being bumped against the rig legs) are needed, together with skilful captains to pilot them. In the North Sea these boats are usually about 160 feet to 250 feet long and are able to carry between 400 and 500 tons of cargo. Normally one rig is serviced by at least one boat, with a second sometimes on hand.

For reasons of speed, helicopters have become the main vehicle for carrying men and lighter supplies to the North Sea rigs; all the latest rigs have a helicopter pad. Apart from being faster the helicopters have also proved safer; transferring crew from a bucking ship can be a

treacherous operation. But fog is the enemy, and there are some days when the helicopter crews are grounded. At a rental charge of £500 an hour it is an expensive inconvenience for the oil companies.

Above A supply ship in heavy seas on a delivery run to a drilling platform.

6. Life on a Rig

THE NOISE ON A RIG is deafening. The roar of the waves and the howling of the wind seem to be drowned by the whine of the drilling operations. The decks are usually wet and slippery, and the air is heavy with the smell of oil, grease and cement. Add to this some of the roughest seas in the world and one can see that life on a rig is anything but a joyride.

The rig is many things. It has been described as "not quite a construction camp, space station or cruise ship," but possessing elements of all three.

It is like a space station because it is relatively isolated; also, because it is packed with the gadgetry of marine engineering. It is like a cruise ship because the cabins are small but comfortable, and the food is good. It is like a construction camp because the work can be rough and hard.

A rig is in operation twenty-four hours a day. The crew normally work a 56 or 60 hour week, with two weeks on and two weeks off. The lowest-paid worker – the labourer, or "roustabout," who does the odd jobs like painting and removing rust – earns around £2,700 a year (1975). The "roughneck" who does the more skilled work on the platform floor can earn around £3,300 a year. The "mud men" who supervise the mud as it is pumped down the drilling string earn over £4,000 a year. The rig boss can earn as much as £10,000 a year.

There are more than 2,000 men at work in the tough, isolated world of the North Sea oil rigs. Everything is

Opposite page A messy job: drilling mud spewing from the drilling string.

provided but there is nothing to spend money on. A skilled worker can go ashore after a standard spell of duty plus overtime with £1,000 or more in his bank account.

Accommodation – often four bunks to a cabin – is free, and so too is the transport home. Below deck the consumption of alcohol is strictly controlled. Men are allowed two cans of beer a day, but they cannot save the ration. There are weekly filmshows, card playing and dartboards, but otherwise recreation is limited. Television reception is possible on rigs fairly near shore, but for those farther out the only real links with the outside world are the wireless and the service boats which bring in the supplies.

Below The mess room on board BP's drilling platform *Sea Quest*.

Rigs are vulnerable to delays of supplies because of bad weather, and can suffer from supply shortages and industrial disputes on shore, which not only hold up the work but add to the rapidly rising costs. Contact with the shore is often difficult. Nowhere else in the world are rigs or platforms faced with such tough conditions for transmitting and receiving wireless messages. Yet rapid communications are vital to safety and to the whole operation. Not only are offshore installations furthest from the coast, in deep water and subject to storms and gales, but their position between Europe, the U.K. and Scandinavia means that they suffer from the shortage of radio frequencies in that area; fishing fleets and merchant ships claim radio priority. Scientists are working on new communication systems to solve this problem.

Work on board a rig can be very monotonous. Much of it is routine. It can be messy, with the decks swilling with water and mud, and dangerous: walking on the slippery walkways around the sides as the rig is heaving in a strong gale demands men with guts (and good sea legs). But it is the tedium which many, especially the roustabouts, remember: the long hours spent painting, scraping rust or washing down the decks. One roustabout found that work for him meant "up to six hours at a stretch hosing down the rig sides and equipment to prevent fire . . . My job soon became an endless succession of doing everything at the double, making work where there was none, painting blue a pipe which had been painted white the day before, and endlessly keeping the encroaching rust at bay."

Above A dangerous job: climbing the drilling derrick.

Within hailing distance of every rig hovers the standby vessel ready to pick up the crew if the rig runs into serious danger of capsizing or sinking. It can also help out if a diver runs into trouble.

Diving is probably the most dangerous of all the rig jobs, and it can bring big rewards. A diver may collect £800 or £1,000 for ten days of very intensive diving. Divers are needed by oil companies on exploration rigs and the production platforms to provide a "first aid"

Above A diver in a decompression chamber after surfacing from deep diving operations.

service. They go down if anything goes wrong. Since it costs £30,000 a day to keep a rig going they have to analyse and solve a problem quickly. Divers are also used on platform construction and pipeline laying. All the time they are working in total darkness and swirling mud. At great depths they become virtual prisoners of the sea.

Below two hundred feet divers work in a diving bell or submersible, breathing a mixture of helium and oxygen. This is known as "saturation" diving, because the helium/oxygen mixture saturates the diver's blood and body to the point at which it would take three days of decompression to bring him back to normal. No matter how long he stays under or is "saturated" with the mix it will still take three days to decompress him. Obviously for a man to work one day and then spend three days being decompressed does not make financial sense, so the divers live at the same pressure in special decompression chambers for as much as three weeks at a time. Often, if they are providing a support service for a rig, much of their time is spent hanging around waiting. Being confined in a small cell for days on end is not everyone's idea of a great job, so it is hardly surprising that the wages are very high. As a result there is no shortage of men wanting to be divers, despite the risks and drawbacks.

Industrial troubles on board the rigs have been rare. Casual signings-on, especially at the roustabout level, have made union organization difficult, although the unions are trying. The Americans in the past have occasionally sacked a whole crew if they sniffed trouble ahead. This is one reason why there has only been one strike in the last four years of North Sea drilling.

Most companies agree that there should be greater trade union involvement, and there are now union members on board rigs. One problem is demarcation between, say, transport workers and boilermakers. Although it is likely to prove a slow job, unionization in North Sea operations is likely to grow.

For many, the reasons for working on a rig are varied, and are not just to do with the money. One roustabout has summed it up: "No-one joins this rig for the money. We all have other reasons ... domestic trouble, boredom with dead-end jobs ashore . . . you name it and someone here has it . . ." The crews are tough and professional. Their lives depend on these qualities.

Above A diver going down to clear the propeller of a tug, caught in a 5-inch tow rope.

BP
BP

7. Dropping the Platform

"WHAT THEY WERE trying to do was drop something into the sea the size of the Post Office Tower and to be sure that this huge platform would land right way up and in the correct position subsequently to drill oil for the rest of Great Britain." (Broadcaster Jack de Manio's account of the "launching" of the Highland One production platform in the Forties Field.)

So you have found a field with enough oil under it to make it worth bringing it ashore. What then? You drop a production platform on top of it which can extract oil from up to thirty oil wells which fan out beneath it. You pump the oil through a pipeline to shore or onto a tanker and your problem is solved.

It is not quite so simple. The oil production platforms for the North Sea are giants by any standards. Highland One which – in the summer of 1974 – was one of the first two production platforms to settle on the North Sea bed (the other being Greythorp One), was described by one observer as: ". . . designed to withstand waves as high as about six cricket pitches and to stand firm in winds of 130 m.p.h. She was made by about 1,700 men from enough steel to build a second Forth Railway Bridge and still leave some over for a railway engine." He might have added that Highland One took two years to build.

The main part of a steel production platform is the steel support structure. This is known as the "jacket." It supports the pre-fabricated deck module which, when erected on top of the jacket, carries crew, equipment and

Opposite page The size of a steel production platform compared with Tower Bridge in London. The platform is 690ft high and weighs 57,000 tons.

Dropping a platform
Left the steel production platform *Graythorp I* during the last stages of construction.

Below The platform being towed out to the Forties oilfield.

Opposite page As the tanks flood the platform sinks into position on the sea bed.

materials. The jacket is built on a large graving dock on the coast. When completed it can weigh up to 20,000 tons. It is then ready to be towed on a specially constructed barge to a spot selected far out at sea. This site has already been determined on the basis of geological information supplied from survey ships.

Towing out and positioning the jacket is a major task in itself. It is essential that the weather is fine – a storm could upset the whole operation. In fact the "dropping" of Highland One in the summer of 1974 was almost delayed because of fears about weather conditions. So a mass of detail has to be collected before a platform is put to sea. Apart from detailed reports from the oil company's meteorological and oceanography experts there are regular reports from weather satellites, weather ships and exploration rigs throughout the North Sea.

The area where the jacket is to be dropped is marked by buoys. In a highly complicated manoeuvre the dropping procedure begins. The barge or flotation raft is gradually filled with water, and the platform tilts to a 45-degree angle. This process, which takes about an hour, is critical. One false move and £60 million of equipment would go plunging into the deep. The "jacket," still on the barge, is then dropped onto the sea bed in a computer-controlled movement which takes ninety seconds. The barge is removed, and over the next few weeks the jacket is fixed securely to the sea bed by massive steel piles. A pre-fabricated deck module, carrying all the materials for drawing the oil from the wells and buildings to house crew and equipment, is then put together on top of the jacket.

The first production platforms have been made of steel – 49,000 tons of it – but the oil companies are now also building concrete production platforms. The first of these have been completed and tested. The industry is divided on the relative merits of steel and concrete platforms. Steel is obviously more likely to corrode, and requires more labour to build. But there are no hard and fast rules for deciding if a steel or concrete platform is more suitable for a particular field.

PROFILE OF A PLATFORM

The giant concrete gravity production platform being built by McAlpine – Sea Tank in Scotland for Shell-Esso's Brent offshore oilfield, 100 miles N.E. of the Shetland Islands in the North Sea.

SOME FACTS:

A cellular base 298 feet square and 177 feet high will rest on the seabed, with four towers rising 352 feet from the base to support a steel deck with an area of 42,000 square feet and weighing 3,100 tons.
Production equipment and living quarters will place a load of more than 6,000 tons on the operational deck. The structure will displace 436,300 tons of water.
It will provide storage for more than 600,000 barrels of crude oil.
Construction to deck level will require
● 257,000 tons of concrete ● 3,100 tons of steel for the deck ● 15,000 tons of reinforcing steel

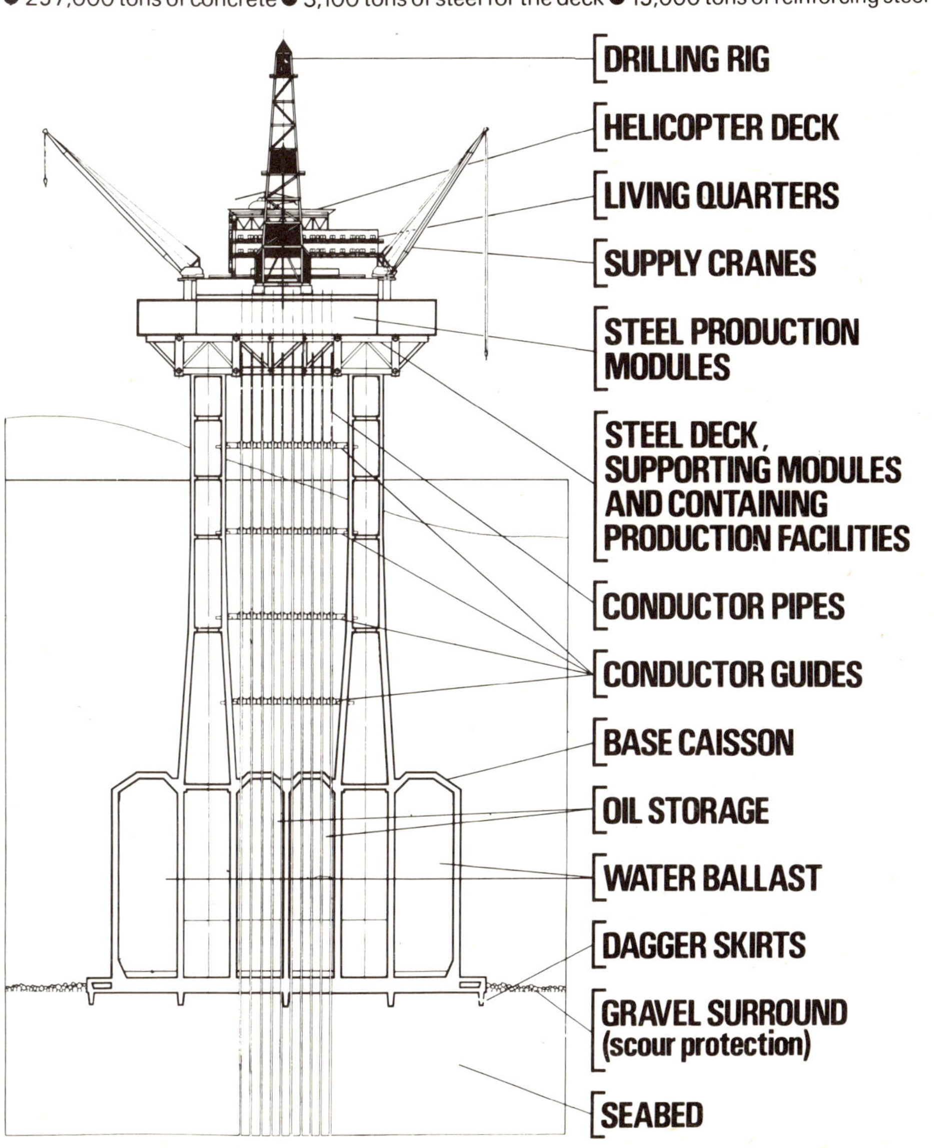

The concrete platform is built upon a base of 19 enormous drums or "cells," each 66 feet wide and 185 feet high. This structure is towed out into deep water offshore where more cells are built on top as support columns and for storing oil. As the structure is built up, it is towed into deeper water and submerged. Built up rather like a giant Meccano set, the completed structure is topped by two steel decks covering 40,000 square feet and accommodating 120 people. Weighing more than 175,000 tons and standing higher than the Post Office

Below The cells of a concrete production platform being built on dry land.

Tower (550 feet), it is finally settled on the sea bed.

The concrete platform is no cheaper to build. Also, concrete platforms have to be built on flat land on the coast and then later in deep water close to the shore. Some are being built on the Clyde, the scene of ship-building for many years. Others are to be built on the east coast of Scotland, in areas that are suitable for the job, but which are also areas of great beauty. This is the cause of the battle between the oil men and the environmentalists.

Below The completed base of the platform being towed out to sea.

SAIPEM-CASTORO II

8. Bringing the Oil Ashore

MOST NORTH SEA OIL will be piped ashore through pipelines under the sea bed. This is the method used for natural gas which, at present, flows through seven pipelines under the North Sea. By the autumn of 1974, British Petroleum had completed its £95 million pipeline which stretches 115 miles between the Forties Field and Cruden Bay, north of Aberdeen. Other pipelines include the Ekofisk to Teeside and the Piper Field line to Flotta on the Orkney Islands, expected to open in 1975. A pipeline taking the oil from the massive Brent field to Sullom Voe in the Shetlands is due for completion in 1976–77. By 1980 some 2,000 miles of pipeline will have been laid in the British section of the North Sea at a cost averaging £600,000 a mile.

The pipelines are specially designed to withstand corrosion on the sea bed. The normal field pipeline, for example, is made from steel pipe of between 20 inches and 30 inches in diameter surrounded by fibreglass and enamel and coated with up to five inches of concrete. It is usually laid from a specially designed ship which feeds it slowly overboard. It may stretch for hundreds of miles. Because of the unevenness of the sea bed, it does not always go in a straight line (although the pipeline layers try to run it as straight as possible). Sometimes deep submarine trenches or rocky outcrops force it to be diverted.

Laying the pipeline calls for great skill. The sections of pipe are gradually pieced together and let out from the

Opposite page The pipe-laying barge *Castoro II* in operation in the North Sea. The pipeline can be seen emerging from the ship at the bottom of the picture.

PIPE BURYING

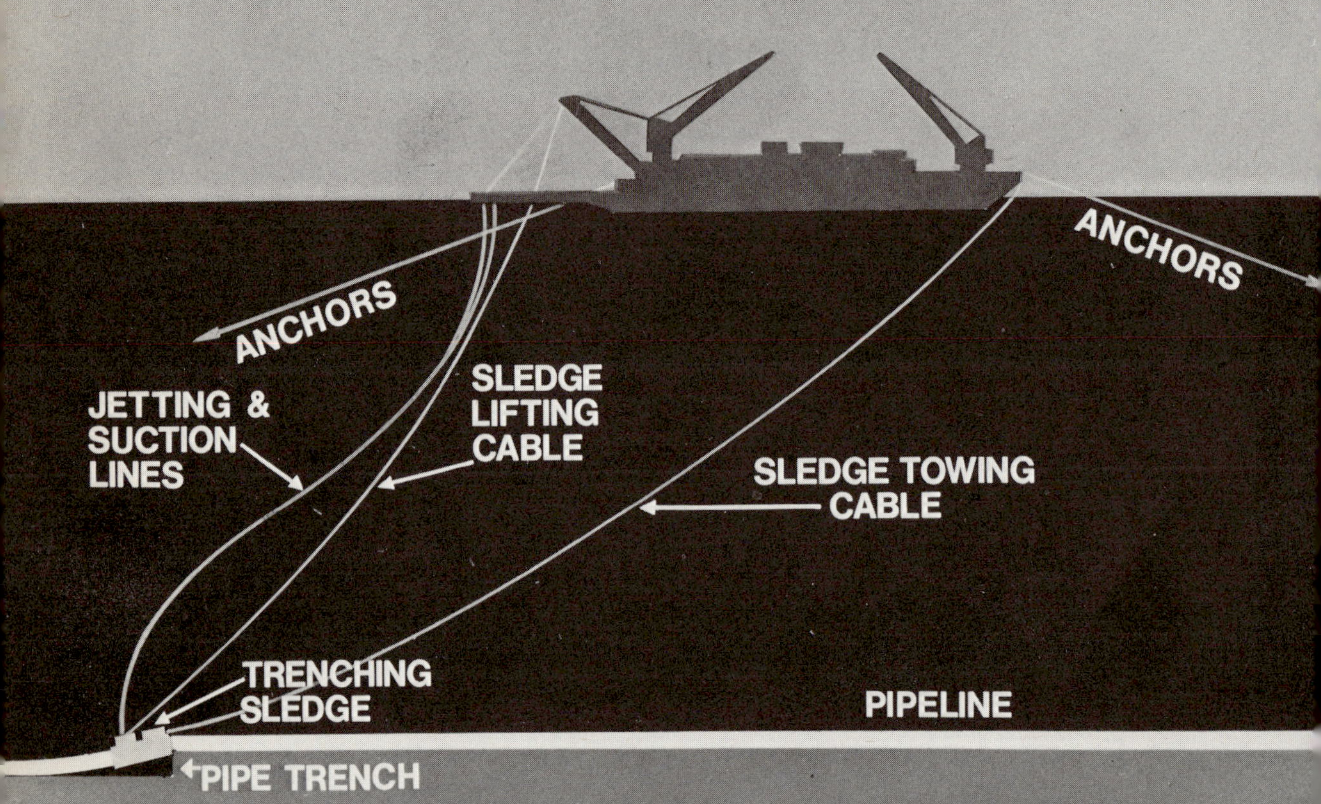

ship, rather as a fisherman on a boat lets out his line. Energetic attempts are made to bury the line in a narrow, shallow trench to stop it buckling under pressure from undersea currents.

Divers must be prepared to go under the sea to ensure that the supply can be kept going, no matter how rough the weather is on the surface. Repairs are few, which is just as well, as conditions for the divers are highly dangerous.

In some cases pipelines cannot be laid, and the oil has to be taken off the rig by tanker. This is necessary in the Norwegian Ekofisk section because of a deep underwater trench which runs a few miles off the Norwegian coast – an impossible barrier for a pipeline to straddle. The tanker docks at specially constructed storage tanks at sea. Yet bad weather can still lead to delays, and it is impossible to have the continuity of supply provided by the pipeline.

On shore, the oil is fed into storage tanks at special terminals. There is a terminal, for example, at Flotta, a tiny island in Scapa Flow at the northerly tip of Scotland. This terminal, which covers 240 acres, will receive oil by pipeline from the Piper Field 125 miles to the east. As well as three conventional tanker landing stages, there will be three large moorings 10,000 feet off shore.

Once on shore the oil is either taken to refineries in other parts of the country or to Europe by tanker, or, as in the case of oil landed at the base near Peterhead, carried in a pipeline on land to the massive oil refinery at Grangemouth, one hundred miles to the south, near Edinburgh.

Above Cutting a trench for BP's pipeline to carry oil from Cruden Bay, Aberdeenshire, to Grangemouth Refinery, on the south shore of the Firth of Forth.

Opposite page A diagram showing how the pipeline, already laid, is buried underneath the sea bed.

9. The Impact Onshore

THE NORTH SEA REVOLUTION has already left its mark on Scotland. On the north shore of the Cromarty Firth around Nigg Bay it can be seen in the huge graving dock on which the first Highland One, one of the first monster steel platforms, was built. To the south, in the cocktail bars of the Aberdeen hotels where block bookings for rooms are kept by the oil companies, it is heard in the accents of Texas and Glasgow, California, Marseilles and Teesside.

Lorries laden with materials for the rigs and the other paraphernalia of the oil world roar along roads once used only by farmers and tourists. Oil supply boats jostle with the old trawlers along the coasts of Aberdeenshire. Remote lochs, such as Loch Hourn, Loch Erribol, Loch Carron and Loch Broom, have been mapped, surveyed, drilled and plumbed – here, too, may be new sources of oil.

In the last few years, service vessels have started operating out of Dundee, Montrose, Peterhead, Leith and Lerwick. New businesses supplying and manufacturing for the oil industry have sprung up around Aberdeen. Rigs are being built on Clydebank, once the home of the great Clydesider ocean liners and battleships. Helicopters bound for the North Sea now take off regularly from Dyce or half a dozen other airports around the north-east of Scotland. A site for a giant concrete platform is now being cleared around the waters of Loch Kishorn in the Western Highlands.

And there is more on the way. In Sullom Voe, on the Shetland Islands, there are plans for a pipeline terminal, tank farm and tanker berth to handle the output from the massive Brent field. A refinery and petrochemical plant is proposed for Loch Erribol as well as a construction site for the building of rigs and production platforms. A pipeline landfall at Peterhead, a construction site near Glasgow and another at Ardesier are all planned. Yet more construction sites are proposed at Burntisland, Evanton and Hunterston. Plans are also afoot for a supply base at Scrabster and a refinery as well

Below The building of a steel production platform on the peaceful shores of a Scottish loch.

as the construction site at Nigg Bay in Inverness-shire. And there are countless smaller plans for pipeline coating plants, servicing areas and refineries.

The oil boom has had a great effect on the Scottish economy. It has already brought hundreds of new jobs to the north-east in particular. The ports and harbours of east Scotland are booming. In 1974 there were around two hundred ships ferrying between the rigs and the Scottish coast. Even on the west coast the proposals to build concrete platforms in the deep lochs and the construction sites and associated industries proposed around Glasgow will bring new jobs and higher incomes to the declining industrial west. The opportunities are there for British companies in manufacturing and servicing.

Even six or seven years ago the scope for industrial change on this scale would have been unthinkable. Scotland had suffered relative economic decline since before the Second World War. The unemployment rate had stayed at twice that of most of Great Britain. The economic growth rate was lower. Too great a dependence on the old industries of shipbuilding, steel making and coal left Scotland vulnerable to economic storms. True, there had been a gradual change. There was the development of new towns like East Kilbride and Cumbernauld, the expansion of light industries and some large projects like the Forth road bridge, Hunterston power station, and the Linwood or Bathgate car plants. But the oil boom has created quite new and exciting opportunities.

In 1974 more people were migrating to Scotland than were leaving the country for the first time in forty years. Latest figures showed a net increase in population of 5,000 compared with a decrease of 10,000 in 1973 – a figure which has been repeated year after year for almost every year since the 1930s. The Scottish wanderer is returning. Samuel Johnson's taunt that the "noblest prospect which a Scotchman ever sees is the high road that leads him to England!" is being turned on its head.

The opportunities are many. For example, in the five

years between 1975 and 1980 between 55 and 80 concrete platforms could be needed. Each platform costs around £60 million and employs about 500 men for two years. This project alone could mean orders worth between £3,000 million and £4,000 million, and employment for up to 5,000.

Although the oil industry produces fewer jobs in proportion to a huge capital outlay than other industries, it still has a great effect on local conditions. Already the impact on the Scottish economy is becoming clear. More than five hundred industrial groups in Scotland are now thought to be involved in one way or another in North Sea activities. One conservative guess is that this could be worth £300 million a year to Scotland's economy from now until the early 1980s. This development has been concentrated in the east and north-east and away from the traditional industrial belt of west and central Scotland. Latest figures from Aberdeen University suggest that two-thirds of the 15,000 jobs already created by North Sea oil, and nearly three-quarters of the 22,000 jobs likely to be created, will be located outside the traditional industrial areas.

But all this expansion is creating conflicts with the existing communities. There is a social price to pay.

10. Problems Onshore

AT FIRST, the Scots welcomed the oil men with open arms. Oil meant jobs which meant more money and a better life. But today they are less certain. For the oil boom has also meant a threat to established communities, the influx of strangers and an explosion in land and house prices. Plans to build rigs or concrete platforms in what, by economic and technical standards, were the best areas, have run into increasing opposition from the local people.

Take Drumbuie, for example. Drumbuie is a tiny village, at the seaward end of Loch Carron. Barely 20 or 30 people live there. They lead a hard life, crofting and fishing for a living, and apart from summer visitors touring one of the loveliest parts of the Western Highlands, few outsiders have bothered them for many years. The young have tended to drift off to the cities for jobs, and never come back. This is the traditional pattern of the area. A hundred years ago the population of the Scottish Highlands was 371,000. Today, just a little over 280,000 (5 per cent of the Scottish population) live there. The drift away has been persistent. But Drumbuie also has good, flat land near the sea. It has a natural harbour, Port Cam, and deep water – perfect for building a concrete platform. A major company applied to build but it faced opposition which grew in strength. After a lengthy public enquiry, the proposals were finally rejected. In other areas, of course, proposals have gone through.

The oil platform builders were rejected from Drumbuie but have been allowed to settle at Loch Kishorn, on the opposite shore of Loch Carron in Wester Ross. At Ardyne Point on the Clyde, where a consortium has invested more than £5 million in a huge platform yard, the opposition was less vocal. After all, this is the Glasgow area, with some of the worst enduring unemployment in the country. It is ironic that the oil boom should have been in the East, which needs such prosperity slightly less than the West.

Industry has fitted in best in areas like Ardesier, which is fairly close to Inverness and Nairn but still reasonably isolated. It has obviously jarred most in the West Highlands, where the few spots around the British coasts capable of taking platforms exist.

Behind the opposition of the people of Drumbuie lay more than a desire to keep life as it is or to preserve a beauty spot. At Drumbuie, and elsewhere in Scotland, there were fears that once the big boom in construction is over, the areas will be left with large yards and construction sites lying empty, having experienced only a five-year burst of life. Also, as skilled labour is scarce, and welders and other skilled workers have to be imported from outside the area, the local people feared disruption to their lives, especially if the construction workers lived in camps.

This was a problem at the Nigg Bay site, for example. By the middle of 1973 the Nigg yard was employing 1,300 men. At first the workers came from Easter Ross, Caithness, Orkney and Sutherland. But as the need for skilled men grew, they came from Glasgow and Edinburgh, South Wales and the North of England. A sudden and heavy burden was put on the slim resources of a county like Easter Ross. Housing was short and the massive influx led to sharp rises in the rents of cottages and houses for 30 or 40 miles around.

Wages were good. A welder or fabricator earned a basic 91p an hour. There was plenty of overtime and "condition money" – an extra 2p an hour for working in mud, 12p for working inside the construction pipes, 10p

Below The slogan of the Scottish Nationalists: "It's Scotland's Oil."

for working over 200 feet and so on. But the shortage of houses meant a rash of caravan sites round Nigg Bay. Temporary accommodation was fixed up in a couple of ageing Greek liners docked in the bay. There was little to do in the evening except drink, and the whole place had some of the atmosphere of the Klondike frontier. This led to some discontent among the new workers which flared up into industrial troubles. From time to time there were walkouts, strikes and picketing. Some of it was due to bad conditions, but much of it was caused by friction between the American employers, unused to British labour conditions, and experienced union men.

Little wonder that some areas, learning from Nigg, set some stringent terms – restoration of the site to its former state, no labour camps or labour ships, local people to get the first offer of jobs and no Sunday work except on essential maintenance (a crucial addition in certain areas of the West Highlands where Sunday is strictly observed as a day of rest).

Yet for all the protests it could be argued that Easter Ross and the Western Highlands have barely been scratched by industry wanting its deep, sheltered waters. There is always a balance to be struck between the demands of the environment and the advantages of new developments, despite the strains they bring.

Towns, too, have felt the strain of catering for the new developments. Aberdeen, the granite city of the North, has become the oil capital of the North. It is the fastest growing town in Britain. With oil has come the doubling of office space. Large areas of new industrial property have been added. In the last four years no fewer than 280 firms have arrived in the Aberdeen area, providing 5,500 new jobs. This has stimulated employment in the service industries catering for the oil industry, such as hotels and transport. Ironically, this boom has created a labour shortage in an area where, not so very long ago, unemployment was high.

The oil bonanza also came at a time when the older industries of North East Scotland – fishing, agriculture, food processing, paper, textiles and engineering – were already enjoying a period of hectic activity. As the winter of 1974 approached, and unemployment throughout Britain was showing an upward trend, Aberdeen had an unemployment rate of 1.4 per cent, against 4.1 per cent for Scotland as a whole. Now the shortage of labour is threatening to hold back North Sea developments.

The Shetland Islands is another area of labour shortages. The situation there is different from Aberdeen; the Shetland Islanders complain that their traditional industries of fishing and fish processing are suffering from a glut in the main market of North America. Lower demand by customers in Europe has hit the knitwear industries. But, at the same time, there are four oil servicing operations being kept busy, and local builders and hauliers have plenty of work. As in Aberdeen and other parts of the North East, the service industry is faced with a shortage of labour. For example, hotels in Lerwick in the Shetland Islands are finding it difficult to recruit staff, shops and bakeries are closing down and motor mechanics and television repairers are hard to find.

The obvious way out is to attract workers from other areas. But here housing and the demand on other services become vital. The strain of the sudden burst of new activity has been immense. Private house prices in Aberdeen match those in the more populous and traditionally more expensive South East of England. Aberdeen County Council used to build about 500 new houses a year. Its future programme has been stepped up to between 4,000 and 5,000 a year, and the amount set aside for housing in 1974–75 was expected to be £20 million. This will place a great strain on the council's finances and much of the burden will fall on the rates. Some estimate that Aberdeen alone needs a further 16,000 houses. Despite the current efforts to expand there is a danger that the shortage of workers and housing in parts of Scotland at the centre of oil developments could hamper progress.

The oil boom has also stretched the road systems which were not designed for the amount of traffic they

Above Sir John Howard, joint chairman of the construction company Howard Doris, points out the site for building a concrete production platform at Loch Kishorn.

Right The neat filling-in of a pipeline trench in Perthshire.

are now taking. Some roads are being modernized, like the A9 between Perth and Invergordon, at a cost of £100 million. Improvements are also being made at airports in the North and East. But there is still little progress on roads like the A94 Perth to Aberdeen road and some others where traffic bearing oil-related goods has doubled or trebled in the last two or three years.

Only now is it being seen just what a burden can be placed on roads, schools and services like the police when an area experiences a sudden boom. In the case of Aberdeen it is predicted that the population will increase by 50,000 over the next 10 to 15 years. Cooperation between local and central authorities is now closer but there is still a long way to go. Only in 1974 did the transfer of government responsibilities for oil activities from London to Scotland begin in earnest. Here

the pressure of the Scottish Nationalists played a part. In the election of February 1974 the Scottish Nationalist Party gained 7 seats, and they raised this to 11 in the October election, taking 30 per cent of the vote. The S.N.P. campaigned under the slogan "It's Scotland's oil," and this played a big part in their success. Their programme on oil and the devolution of power is rather vague, and it is doubtful how many Scots would want complete separation from England; but certainly S.N.P. pressure has encouraged moves like those outlined in the Government White Paper of September 1974. Its authors suggest a Scottish Assembly which would receive an allocation of at least some of the oil revenue. The future, both political and economic, is likely to reflect the new force of political nationalism encouraged in part by the extraction of North Sea oil.

Above Loch Kishorn – where a concrete production platform is being built.

11. Costs You Can Measure

THE FINANCIAL COST of developing North Sea oil is sky-high. Every day of the year some £2 million is being spent. By 1980, if the oil is to flow in the quantity needed, £10,000 million will have to be found. If the oil industry is to bring ashore the four million barrels a day it expects by the mid-1980s the total development costs could top £25,000 million at 1975 prices.

In the decade to the end of 1974 nearly £1,500 million had already been spent in the North Sea. About 700 appraisal and production wells had been drilled at a cost of around £1 million each. Winter drilling in northern waters, where it is possible, can cost between £2–£3 million a well.

Costs vary from one area to another, and seem to rise all the time. Inflation is hitting hard. When B.P. began work on its Forties Field it expected to have spent £350 million (£831 per barrel) by the time it came on to full stream, producing 400,000 barrels a day in 1976. This estimate has now soared to £600 million (£1,500 per barrel) a day. The Ninian field off the north-east Shetlands is likely to cost between £800 million and £1,000 million to develop. The latest development charge of a new oil field in the North Sea is now put at £3,000 per daily barrel. Only a year ago it would have been £1,000.

The oil company trying to work out whether a field is worth developing or not must make some careful calculations about the cost. Here is a check list of some

Opposite page An expensive job: laying a pipeline under the North Sea costs about £600,000 for every mile.

of the major items:

Rigs	£6,000–£10,000 a day to hire
Production platforms	Steel: £40 million+
	Concrete: £60 million+
Offshore services (catering, diving, etc.)	£2,500 a day
Insurance	£1,000 a day
Wages	£2,500 a day
Helicopters, workboats	£3,000 a day
Materials (well casings, mud, cement, etc.)	£6,000 per well
Pipeline	Around £600,000 a mile

The costs of exploiting a field of 300,000 barrels a day, even in 300 feet of water, would be something like those shown on the following list (indicated in millions of pounds):

	Approximate cost (in £m.)
Drilling five preliminary wells	10
Drilling rigs (five fixed)	150
Production platform (one concrete)	60
Insurance	18
Hire of drilling rig	12
Wages (1,800 operational days)	8
Helicopters, workboats, etc.	15
Materials (35 wells drilled)	10
Pipeline (120 miles)	72

The total cost would be around £355 million, or £1,150 per barrel per day. This is a simple example in not very deep water by North Sea standards, and, as B.P.'s experience in the Forties Field shows, inflation and rising costs are likely to push this figure up. These vast sums of money have to be raised by the companies or the Government depending on the arrangements in each particular field. As the oil begins to flow the costs, hopefully, are recovered. It is with the hope of huge profits to come that the companies are prepared to shoulder the gigantic costs.

12. Financing North Sea Oil

B.P.'S FORTIES FIELD is now likely to cost twice the £350 million originally expected. For every 1,000 barrels a day of production from the North Sea, around £1,500 will have to be found to bring it out. If you add the cost of onshore activities, for every £1 spent directly in the North Sea £3 could be spent ashore. At present, only a third of the development money in the North Sea is being supplied by British companies. By 1980, however, this should be up to around three-quarters.

Where does all this money come from? Even for the giants, the problems are immense. A decade ago, oil companies used to boast that around 90 per cent of their expenditure was self-financing. In other words, the money earned by their successful wells paid for the exploration and development of new wells. Today, because of the huge sums of money involved, the oil companies can only provide around 40 per cent from their own resources. So they have turned to consortia – a collection of banks, finance companies and Government agencies – for the funds. When British Petroleum raised £370 million some time ago, the money came from a consortium of around sixty banks, many of them American or European.

For the big companies, with a good field and an international reputation, finding the money is not too great a problem, although it can take time. Most of this financing – as in the case of the Forties Field – is what is known as "off balance sheet," or "non recourse." This means that the loans can only be used for the particular oil field for which the money was raised. Other companies have also raised loans on the Eurocurrency market – the vast pool of international funds which can be used for long-term financing. With the Government showing greater interest in taking direct stakes in North Sea oil, the Exchequer may become a further source of funds. As we have seen, some of these are likely to be spent on taking stakes in future developments directly. Even so, most of the huge sums are still likely to come from the private financial sector, although towards the end of 1974 there was much less money available. Moreover the costs of certain basic items rose by between 30 per cent and 100 per cent. Certainly by early 1975 new finance was still being raised. But funds were scarcer and terms stricter. Few people expected any major change until the oil companies knew exactly what they were up against in terms of future taxation and the extent of Government involvement.

Below A graph showing the profit to shareholders from the oil industry over ten years.

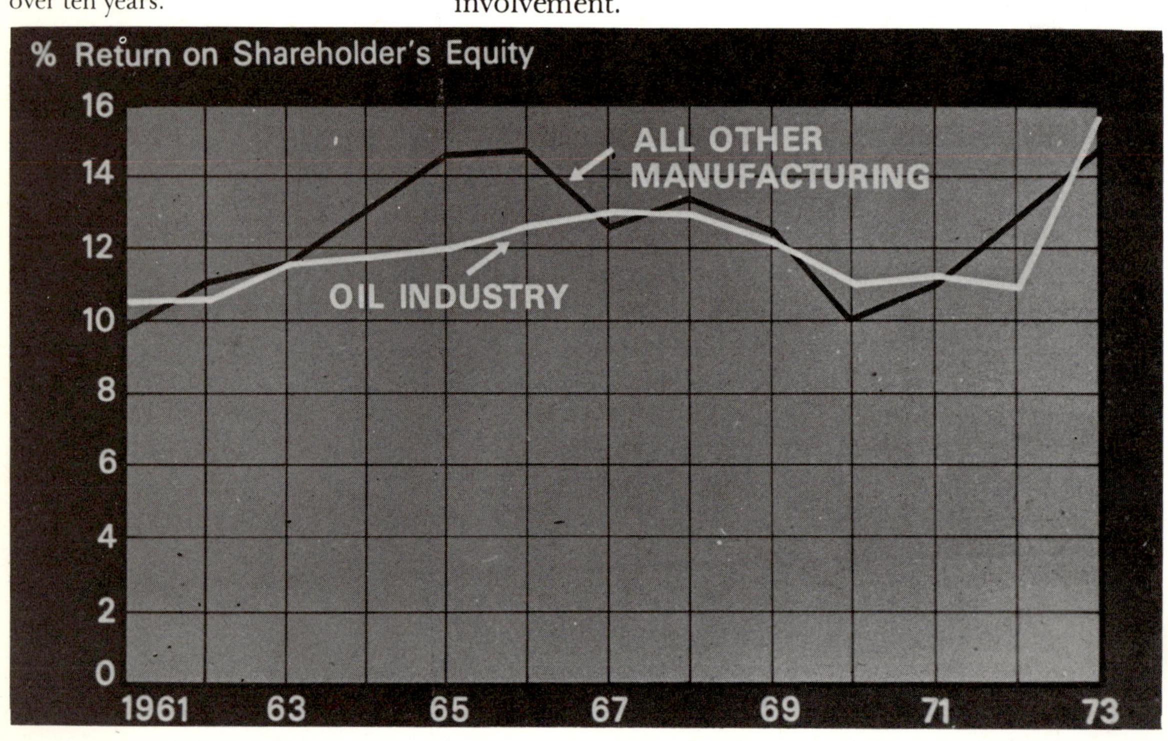

13. The Financial Benefits

WHAT IS THE potential profit from North Sea oil, for the companies and the country? This has become the hardest North Sea question of all, for it depends on so many factors. The key ones are:

1. The amount of oil which is finally recovered from the North Sea: the more oil, the higher the profit.
2. The world market price of oil when North Sea oil starts to flow and the final cost at which the oil is produced.
3. The political attitude to the oil profits in the form of taxation, government involvement in exploration and the government's encouragement (or discouragement) of further exploration in more difficult waters.

First, the oil reserves. No one is sure just how much oil lies beneath the North Sea. In the last year or so the estimates have been revised steadily upwards. From 10,000 million barrels of recoverable oil the figure has risen to 13,000 million, to 18,000 million and (unofficially) by December 1974 to 22,000 million. Some observers believe, however, that the total recoverable reserves in the U.K. sector could reach 50,000 million barrels.

By 1980 the Department of Energy expects production to be in the range of 100 million to 150 million tons a year – more than Britain used in 1973. Some oil men would put the figure higher, at around 170 million tons, or 140 million gallons a day. This would be true if there were no hitches. But there could be plenty of them – and not just the technical snags of delays in building platforms or finding materials. The crippling cost of bringing North Sea fields into operation is a growing burden on even the mightiest companies, and the uncertainty over the British Government's plans for participation in North Sea oil could lead to long delays.

The more cautious experts think that it may prove impossible to bring more than 100 million tons ashore by 1980. By 1985, however, the total could be nearer 200 million. But it must not be forgotten that the oil will not last for ever, and although new discoveries in, say, the waters west of the Shetlands or in the Irish Sea could extend the reserves, present estimates show production reaching a peak in the 1980s and then starting to tail off in the 1990s. This is important not only in viewing the North Sea oil as a national resource but also in considering the profitability of the fields. Peak profits will be made at peak production.

Another problem is that no one is sure what the price of oil will be in 1980. When oil was first found in the North Sea it was priced at around three dollars, or slightly over £1 a barrel. But the massive rise in oil prices forced by the Middle East producers in 1974 has pushed this up to nearer £5 a barrel. This has obviously had a great effect on the profit prospects for the companies, and it is one reason why any British Government, Labour or Conservative, is certain to tax the profits coming from this sharp rise in oil prices.

But there is no agreement on what oil prices could be in 1980. Some experts believe that oil in 1980 will cost about two-thirds more than it does now. But not everyone agrees. Others feel that oil prices in 1980 could be lower in real terms, after allowing for the upward trend in inflation. They say that other oil supplies will have been developed, and that there could even be a surplus of oil. Also, new sources of energy, such as nuclear energy, could be playing a bigger part in our economy. And if people were less dependent on oil, prices would tend to fall. In this case, the financial benefits of the North Sea, although still vital, would not be as large as the most hopeful forecasters expect.

Nevertheless, given the continued demand for oil which is likely in 1980 (as explained in the introductory chapter), the North Sea will still be important, especially to Britain's trade balance. In 1974, for example, the deficit on Britain's balance of payments (which reflects the gap between what we buy from abroad and what we sell) was in the red by about £4,000 million. Some three-quarters of this was due to the "oil deficit" brought about by the vast increase in the price of oil. North Sea oil should make a big impact here. On the estimates of the major oil companies, the North Sea contribution to our balance of payments on the basis of the current price of oil could be as high as £1,400 million by 1976. This could rise to £2,300 million in 1977, and over £3,000 million in 1978–79. On the basis of a flow of half a million tons, or 3.5 million barrels a day, the balance of

Below The Saudi Arabian Oil Minister, Sheik Ahmed Yamani; the Algerian Energy Minister, Mr. Belaid Abdesslan; and Mr. Tom Boardman, the then British Minister for Industry. In November 1973 these three men met to discuss the relationship of Great Britain with the Arab oil-producing countries.

payments contribution of North Sea oil in 1980 would be around £5,000 million, and more than £10,000 million by 1985.

Despite the difficulty of working out the national benefits to the last £1,000 million it is clear that Britain's economic prospects would be a lot bleaker if we did not have North Sea oil to look forward to. Moreover, the biggest advantage to Britain could lie in Britain becoming an oil exporter, something not even thought possible ten years ago. We shall not use all the oil we produce. As explained in the Introduction, North Sea oil is a special type of oil, a light oil which is found in very few areas of the world. (The oil found in the Persian Gulf, for instance, is a heavier kind of oil.) We shall, therefore, be able to export our oil to countries who cannot produce it themselves. On the other hand, we will still have to import the heavy oil from the Middle East, for we have nothing like it in the North Sea. The wheels of industry would seize up if we had to rely on North Sea oil alone to keep them turning. Thus it will assume key importance as a medium of barter.

But, once again, the benefits North Sea oil could bring to the country, as well as to companies involved, will depend on how quickly it is brought to land. In this, the influence of politics could prove crucial.

Below The meeting on 20th August, 1971, to grant permission to oil companies to drill in fifteen areas of the North Sea.

14. Politics and the North Sea

AT TWO O'CLOCK in the afternoon of 20th August, 1971, a few top civil servants started opening some sealed envelopes in room 1247, Thames House South, the London home of what was then the Department of Trade and Industry. The envelopes contained the bids of a number of top oil giants for permission to drill in 15 new 100 square mile blocks of the North Sea. £21 million was offered by Shell for one block alone – block 211/11. This sale by tender raised £37 million for the Exchequer.

This "auction," hardly noticed at the time, is a reminder of two facts often forgotten. Firstly, the oil game is a risk. (Shell, who put up that £21 million for its "golden block," has found it depressingly dry.) Secondly, the Government has been involved in control of North Sea developments for some time. Once it became obvious that the oil companies were becoming seriously interested in the North Sea, the Government of the day made the first moves to ensure it could keep an eye on things. The Continental Shelf Act (1964) gave all mineral rights in the British sector of the North Sea to the Crown. Also, it set down various regulations about the licence to explore in British waters. For one thing, companies would have to be registered in Britain. Half the licence would have to be surrendered to the Government after six years. The terms on which the rest could be kept had to be re-negotiated on a forty year lease with a steadily rising rental. Royalties in favour of the

Government were fixed at $12\frac{1}{2}$ per cent. Each licence also reserved the right of the Government to move into a field and bring out the oil itself if it was not happy about the rate at which it was being developed. Right from the start the Government was involved.

The first three rounds of licences – in 1964, 1965 and 1969 – were allocated by the Government. The fourth – in 1971 – was put up to auction.

Before the first round of allocations, the British area of the North Sea was divided into blocks of about 100 square miles each. The Government then invited the oil companies to request permission to explore whichever blocks they thought worthwhile. Most of the companies had already made initial surveys of the North Sea, and for the more promising blocks there were several applications. In order to be fair to all, the Government split up the best areas among different companies. As a result, everyone had a chance of "striking it rich."

At the moment direct British participation in North Sea licences is 30 per cent, with the stake held by State enterprises like the National Coal Board and the Gas Council around an eighth of the total. But because there was some discretion in allocating licences, the British share of the oil found is as high as 45 per cent.

Certainly, the North Sea is an international attraction. Of the 250-odd companies involved in exploration alone, 100 are American, 75 are British and the rest include Canadians, French and Germans. Apart from British giants British Petroleum (in which the British Government has owned a 48.2 per cent stake for years), Burmah and Shell, there are groups like America's Occidental, Esso and Conoco, France's Total and a number of non-oil groups such as the Bank of Scotland, Rio Tinto Zinc and Associated Newspapers.

As the full extent of North Sea oil reserves became clear all political parties realized that the share of possible riches going to the oil companies, who bore most of the risk, was too high. From 1973 onwards there was a debate, still far from being ended, about the best ways to improve the Government's "take" of North Sea oil.

Soaring prices from 1974 gave further impetus to these political moves.

In broad terms the Conservative Government favoured "creaming" the profits off through a special tax, on top of the 12½ per cent royalty and the standard Corporation Tax, levied on all companies at 50 per cent. This was rather like Petroleum Revenue Tax outlined by the Labour Government. But the Labour Government has since gone much further. It has set out its approach in two Bills – the Oil Taxation Bill, published in November 1974, and the Petroleum and Pipelines Bill, which appeared in April 1975. By the summer of 1975 both Bills, amended in parts but unchanged in detail, were well on their way through Parliament to become law.

The Oil Taxation Bill outlined the Government's plan for a Petroleum Revenue Tax, to be levied at 45 per cent. While introducing the Bill, the Government stated that it expected the revenue from North Sea oil between 1975 and 1980 to provide a total income of £4,000 million. By the early 1980s, this could rise to between £2,000 million and £3,000 million.

This provisional calculation was based on an estimate of oil revenue from three sources: royalties at 12½ per cent of well head oil value, Petroleum Revenue Tax at 45 per cent of revenue from sales (after deducting royalties and expenditure) and Corporation Tax at 50 per cent.

The Petroleum and Pipelines Bill revealed the Government's plans for 51 per cent "participation" in commercial oil fields and new fields and the creation of a British National Oil Corporation. The BNOC will have powers to involve itself in many aspects of the oil world, from exploring to selling the product.

The Government's plans outlined in these two Bills have caused a considerable row in the oil industry. American companies, in particular, have been annoyed at proposals which they feel will discourage further exploration. Some of them have already started to move their rigs out of the North Sea to the Gulf of Mexico, and a number have stated that they will not continue their exploration in North Sea waters.

Participation has also been criticized because of its likely cost. Estimates of how much the Government will have to pay to compensate companies for the loss of 51 per cent of a highly profitable asset vary from £1,000 million to £4,000 million. This has come at a time of growing financial crisis at home and abroad, with the Government already raising large loans from wealthy Middle East oil-producing countries against the security of future flows of oil from the North Sea. It is rather like raising a loan on the security of a house. But, unlike the housebuyer, the Government is really making a guess on the value of the North Sea in the years to come. If the price of oil falls the debts will be much harder to pay back. One expert has put the likely North Sea "mortgage" at a massive £30,000 million by 1978, a debt which would wipe out most of the potential benefit.

On other important points, like future exploration and the rate at which North Sea oil is to be used, the Government appears to be taking a softer line. But the whole discussion over the North Sea has caused a delay in the development programme. In the meantime, rising costs and the difficulty of raising cash are already forcing oil companies to change their budgets all over again. After all, the sums involved mean that British industry will have to plough more cash into the development of the North Sea than it has into any other area of industrial development in our history (excluding the money spent on armaments during the First and Second Worlds Wars).

15. Dangers of the North Sea

Oil spills

IN MARCH 1967 the *Torrey Canyon*, a 61,000-ton tanker and one of the largest ships then afloat, smashed into the rocks off the Scilly Islands, off the south-west tip of England. More than 95,000 tons of crude oil spewed out of its tanks into the sea. The oil slick, which stretched for miles, left its ugly tide mark along Britain's coastline for months afterwards. Since then there have been a number of oil disasters: dying, oil-soaked birds or soft balls of oil on holiday beaches have become a reminder that there is another price to pay for the many benefits that oil brings.

Ever since offshore drillings began, environmentalists have been conscious of the new type of danger lurking in an accidental mixture of oil and water. When oil spills on land it spreads slowly and can be controlled quickly. But oil spilled offshore spreads swiftly, is hard to control and can cause terrible damage to marine life and birds. Moreover the chemicals used to fight the polluted seas and beaches can themselves cause harm, especially to marine life. After the *Torrey Canyon* disaster the Royal Society for the Protection of Birds calculated that some 25,000 birds, mainly guillemots and razorbills, had perished as a result of either crude oil or the chemical dispersants.

So far there have been no disasters in the North Sea, but the risk is always there, despite precautions by the oil companies, and recent history elsewhere shows that offshore spills are not impossible. One of the worst oil

Above Oil pouring from the tanker *Torrey Canyon*, wrecked off Land's End on 20th March, 1967. The oil was washed ashore, causing terrible pollution of Britain's beaches.

spillages took place in the Gulf of Mexico in 1971, when a platform drawing oil from twenty-two wells exploded. It took four months to check the seepage at the cost of four lives and 36 million dollars.

In 1969 one of the worst offshore spillages occurred in California when there was a blow-out on a drilling rig in the Santa Barbara channel. A blow-out takes place when the drill digs into areas where the pressure of oil or gas is higher than expected. It shoots up in the old-fashioned "gusher." Although the gusher was brought under control fairly quickly, by then more than three million gallons of oil had poured into the sea and few beaches along the California coast remained untouched.

Following the Santa Barbara disaster, a special Presidential panel was set up which reported that: "As off-shore development continues to expand at the present rate and the frequency of accidents remains the same, we can expect to have a major pollution incident somewhere every year. It is noteworthy that the

Above Oil from the *Torrey Canyon* on a south coast beach.

possibility of oil leakage and gas leakage from off-shore oil production must never be completely discounted for the probability of blow-outs cannot be reduced to zero."

But these spillages must be put in perspective. In 1972, for example, a United States Senate report calculated that between 1938 and 1972 some 13,500 wells were drilled in the Gulf of Mexico. During that thirty-four year period there were only seven spillage incidents. There have only been two incidents in the whole of the Persian Gulf's history.

In reply to their critics the oil companies do not deny the risks but argue that those who criticize offshore developments because of pollution dangers exaggerate the relatively small number of spillages and, much more important, do not understand how pollution can be caused offshore and how it can be prevented.

There are two main sources of pollution danger. The first is during drilling, when a blow-out would occur (as in the Santa Barbara disaster), or through seepage from the oil-bearing strata. The second can occur at the production stage or when the oil is being carried by pipeline or tanker.

In the North Sea, oil companies have sometimes come across gas pressures higher than they expected, which could have led to a blow-out. At the production stage there are always dangers (admittedly remote) of a production platform being battered or smashed into by a tanker. Pipelines could become uncovered by the movement of the sea and prove vulnerable to ships' anchors, or the tides. Moreover, underwater pipelines seem to attract fish in shallow waters and are a tempting area for fishermen to trawl. Divers on the West Sole Gas line have found dozens of trawler boards lodged against the pipeline where it had become exposed to the sea. Any damage to a pipeline could create an oil slick.

Yet over the last decade the oil companies have invented all kinds of safety devices. During drilling the first precaution is the "mud," which among other things acts as a wall between the well and the steel casing which surrounds the drilling string. A complicated system of valves and automatic shut-offs is installed on every rig and production platform to prevent disaster. Shut-off points also operate at all key stages of the loading and pipeline operations. Much of the energy and expense which went into developing extra strong steel and cement for the pipelines was aimed at preventing any risk of pollution. Added to these precautions is the experience of the operators who have been dealing with the problem for years.

Since the *Torrey Canyon* disaster there have been moves to improve the chemical "detergents" used to clean up the oil, to have standby equipment ready, to tighten up insurance cover and control the spill as quickly as possible, partly through various types of "booms" or gates which confine the oil slick. But the problem of controlling oil slicks in rough water is still a headache for the scientists and there is no foolproof system yet.

Above When an oil slick from a passing tanker covered Kent beaches hundreds of seabirds were badly affected. Here an RSPCA Inspector examines dead guillemots, their feathers stuck together with oil.

Much of the oil which spoils our beaches comes not from accidental disasters but from the swilling out of a tanker's tanks with sea water on a return trip to, say, the Persian Gulf. British law states that no ship may discharge oil or oily mixture within three miles from the shore. Beyond this, however, the legal position is unclear. Under an international agreement no discharge is allowed within fifty miles of land anywhere in the world, but some fifty countries have not signed it. These regulations have been tightened up, but international co-operation on this law of the sea is far from strong. Britain recently increased the maximum penalty on firms found guilty of oil pollution from £1,000 to £50,000. This is the strictest penalty in the world.

The large international oil and shipping companies all warn their captains not to infringe the law, and to watch for the dangers stemming from possible pollution accidents. But there are still too many companies and countries which turn a blind eye. As with the dangers of a spill, it is in the interests of both oil companies and governments to prevent an accident. The oil companies face enormous cost and difficulty in restoring production, as well as earning themselves a bad name. The Government would have to cope with the political outcry. Together with the environmental arguments the pressures are on governments and companies constantly to press ahead on limiting any pollution danger, no matter how remote it may seem.

Man overboard

Pollution is not the only danger in the North Sea. In December 1965 severe storms swept the North Sea, and during one of them the exploration rig *Sea Gem* collapsed. Thirteen of the crew were swept overboard and drowned.

Sea Gem has been the single biggest disaster, but there have been smaller ones in which men have lost their lives. In 1974 alone 10 men died in the British sector, bringing to 41 the number of casualties since exploration began in 1965. Apart from the 13 lost on *Sea Gem*, 9 have died on the drilling floor of rigs, 6 in dives, 4 in crane accidents and 9 from other dangers like the whiplash from an anchor chain, falling from the platform of a rig and so on.

Above Diving is one of the most dangerous jobs on an oil rig.

Other accidents have happened without loss of life. The shifting of sand after a storm several years ago led to the sinking of another rig, the *Ocean Prince*, off the Dogger Bank. In 1970 *Constellation* capsized and sank while under tow. More recently, in 1973, the rig *Trans Ocean III* was lost. North Sea exploration has claimed four supply and standby boats besides damaging many others. A storm once tore a French barge from its moorings in Norwegian waters in the Ekofisk field and swept it 250 miles north to the Shetland Islands.

Danger lies under water, too. Divers are essential in the North Sea but they work at depths much greater than normal and, although safety standards are high, accidents have happened. In the summer season, when activity is most hectic, there can be as many as six hundred divers at work in the North Sea, drawn from all over the world by the lure of big money. Diving is a dangerous job anywhere, but the rush for oil has sent men deeper into the North Sea than ever before. Diving to 600 feet – unheard of a few years ago – is now a common occurrence and there are plans to go down well below 1,400 feet.

Long experience has shown that the effects of a diver ascending to the surface too rapidly, even from shallow water, can be fatal. The result is a condition known as "the bends." While they are working at great depths divers breathe a mixture of nitrogen and oxygen, which is dissolved in the blood under high pressure. If a diver returns to the normal pressure of the surface too quickly, the nitrogen in his blood forms tiny bubbles, causing great pain, and sometimes death.

Deep-water diving is throwing up new hazards. Beyond 160 feet it is unsafe to dive in a diving suit, so diving bells have to be used. At these depths the diver exists on a mixture of oxygen and helium (the nitrogen/oxygen mixture used in shallower waters is no good, because at great depths nitrogen brings on "the rapture of the deep" where divers drift into a daze). The long-term effects of "saturation" diving (see Chapter 6) at great depths are still unknown, as are the effects of living for days at a time in small decompression chambers. Moreover, the problems of working at the depths to which divers are now likely to go – 600 to 1,000 feet – are still unfamiliar to medical scientists. Even at 100 feet, for example, a diver will be facing a pressure on his body of 440 lbs. per square inch, compared to 15 lbs. per square inch on the surface. Provided everything works all is well, but any abrupt change in his artificial environment would lead to his swift death.

From 1st January, 1975 new diving regulations were introduced which set medical tests for divers and fixed the maximum hours for which a diver can stay down. Regulations on the rig are not quite so strict, yet the dangers are still there, especially on the drilling floor. The most vulnerable workers are the untrained roustabouts and labourers who lack the experience of

Below A driller is lowered from the drilling bit. Accidents are sometimes caused by men falling off rigs.

the older hands.

The Department of Energy, however, in late 1974 were working on various regulations which are expected to cover the safety and health of workers on the rigs as well as minimum safety requirements. Obviously many of the rigs, especially those of the large companies, are well above any minimum requirement, but as in any rapidly growing business there are exceptions, and it is these the new regulations would be designed to cover.

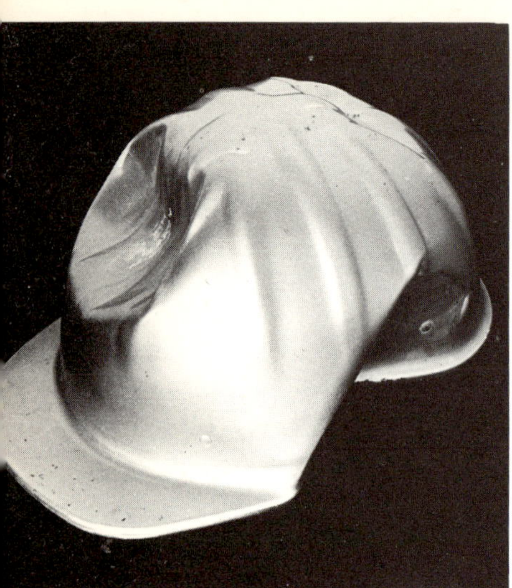

Above This safety helmet saved a man's life; his head was jammed between the drilling pipe and the deck of the rig during stormy weather.

Security

One problem connected with North Sea oil is rarely raised – security. Yet it is of vital importance. By the early 1980s, the North Sea oil and gas supplies will form a crucial part of our national wealth. But are we doing enough to protect it? So far discussions about how to protect our North Sea oil installations from attack or sabotage are at a very early stage. Yet this is the age of the saboteur and the terrorist. An offshore oil platform is a vulnerable and valuable target. So, too, are the undersea pipelines linking the oil well platforms with the onshore storage depots and refineries – especially where they run into the shallow waters near the coast. This is more than the stuff of stirring fiction. By 1980 there will be over 30 production platforms in the North Sea. There will be hundreds of miles of pipeline and many oil depots. A determined guerilla group could create considerable havoc and use blackmail power unless proper steps are taken. These could prove expensive. Any Oil Protection Service would have to have long range aircraft, helicopters based on the platforms and possibly hydrofoils – which are not only fast but also remarkably stable in rough seas.

The security problems of offshore oil are by no means limited to Britain. Many areas of the world are being explored for offshore oil, from the seas of North West Borneo to the Australian coast, the Pacific coast of the U.S.S.R. and the North Aegean Sea. Oil has already become a powerful political weapon. Protecting it is certain to develop into an essential part of national defence.

16. And After the North Sea?

Fresh (oil) fields

EXPLORATION OF THE NORTH SEA is far from over. But already oilmen are looking at other even more inaccessible or difficult areas around Britain's shores. The Celtic Sea is one such area. This is the inland sea basin which includes the Irish Sea, the Welsh coast and the waters off the Cornish peninsula. So far little has been found – only a few traces of gas. The attractions of the fruitful areas of the North Sea have meant that there is no great urgency behind exploration in the Celtic Sea. Nevertheless, it could provide a reservoir of oil for future years. So, too, could the waters in the Minch, to the west of the Hebrides, and the seas to the west of the Shetlands, where the outlook is more promising. Recently, Britain formally annexed Rockall, the lonely, forbidding rock three hundred miles out in the Atlantic, as a part of Inverness-shire – just in case oil is discovered around it.

The outlook, therefore, is of continued exploration with many more oil fields, some large, some small, being discovered. Because the best fields have been tackled first, it seems likely that there will be more dry wells. But if the present demand for oil continues and a great effort is made in developing our own oil so as to cut down our dependence on others', the seas around our coasts will have to provide new oil for many years to come.

New Techniques

Exploring for and developing oil in areas like the West

Shetland basin would be even more hazardous than in the North Sea. This is why the scientists are already working on new techniques for operating in very deep water. At the moment, the steel or concrete production platforms can cope with oil fields lying in water up to four hundred feet deep. But some future oil fields may be found at depths of well over six hundred feet. Existing platforms would be too impractical and expensive. This is why the oil groups are looking at the idea of submarine production platforms. At the heart of this futuristic system is an underwater station on the sea bed, called a satellite. The base of the satellite will act as a circular plate, through which wells can be drilled from drill ships or semi-submersibles. From the satellite, divers will build a production station on the sea bed linked to the surface with sophisticated diving bells or small submarines. The main living and control section will be kept at normal air temperature.

Eventually, large amounts of the world's oil could be produced from these underwater stations lying on the ocean bed hundreds of feet below the surface, but all this could be some fifteen or twenty years ahead.

Below A new development; this concrete production platform has a perforated collar to lessen the shock of powerful waves smashing into it.

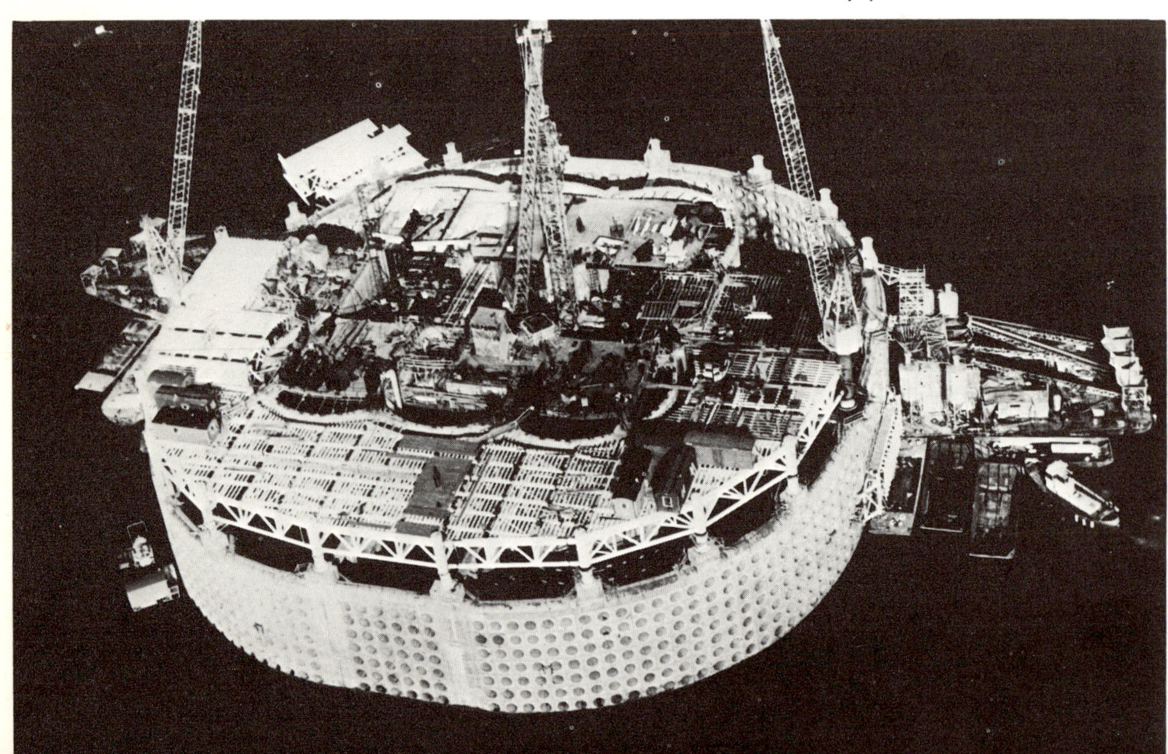

North Sea Timetable

1959, August First North Sea gas discovered near Schlochtern, in the Netherlands.

1964 Continental Shelf Act vests all mineral rights in the British Sector of the Continental Shelf in the Crown.

—, September First exploration licences in the U.K. sector granted. These cover 348 blocks and involved 22 companies.

1965, November Second round of exploration licences granted: 122 more blocks allocated.

—, December B.P. discover West Sole gas field.

1966, April Shell Expro discover the Leman gas field.

1967, March First gas piped ashore from West Sole to Easington in Yorkshire.

1969, June Third licensing round.

—, December Amoco/Gas Council strike a "possible commercial oilfield" 150 miles off Aberdeen.

1970, June Norwegians make "a giant oil find" – Ekofisk and West Ekofisk fields.

—, October B.P. discover U.K.'s first major oil find – The Forties field 120 miles off Aberdeen.

1971, February Shell Expro discover Auk field in Scottish waters.

—, August Fourth round of allocations. Fifteen blocks offered for sealed cash tender attract winning bids worth £37.2 million.

1972, August Shell Expro announce discovery of the Brent field, the most northerly offshore field in the world. Hamilton reveal Argyll field find.

—, September Mobil announce the discovery of the Beryl field north-east of the Orkneys. Cormorant and Thistle fields also discovered.

1973, January Piper and Maureen fields announced.

—, February Dunlin field estimated 250,000 barrels a day.

—, September Conoco announce 200,000 barrels a day from the Hutton field.

1974, February Massive Ninian field revealed 100 miles north-east of Shetland Islands.

—, April Unocal confirm oil discovery in the East Shetlands basin.

—, May Marathon discover a commercial gas field in the Celtic Sea.

—, October B.P. confirms Andrew field north-east of the Moray Firth.

—, November Oil Taxation Bill published by the Labour Government. Talks on participation in North Sea fields start between the Government and oil companies.

—, December Texaco reveal a "significant" new find north-west of the Forties field.

1975, February Texaco announce what could be one of the richest strikes at Block 14/20, near the Piper and Claymore fields. The Government decides that Petroleum Revenue Tax is to be fixed at 45 per cent.

—, April Petroleum and Pipelines Bill published. The Government reduces its forecasts of oil likely in 1975 to between 1 and 3 million barrels.

—, June First North Sea oil comes ashore – 14,000 tons landed by tanker from the Argyll field. *Greythorp I* and *Occidental Oxy*, two more steel platforms, dropped in Forties and Piper fields respectively. First concrete platform, destined for Beryl field, nears completion.

—, August Texaco strike oil and gas reserves in the East Shetland basin.

Glossary

BARREL The standard unit of measurement of oil. One barrel of oil is equivalent to 35 Imperial Gallons, and would produce about 30 gallons of petrol.

CRUDE OIL Oil in its natural state, before it is refined to separate the substances from which it is made up.

DERRICK BARGE A barge carrying one or more derricks, or cranes, used to lift heavy materials in the construction of production platforms.

DIVING BELL A large, airtight container lowered from a ship, in which divers descend to great depths to work on the sea bed.

BIT The tip of the drilling string. It has two or more toothed wheels which cut into the rock beneath the sea bed when the string is rotated from the drilling platform on the rig.

FIXED PLATFORM RIG A rig fixed directly to the sea or river bed. This type of rig is only suitable for use in shallow water.

GRAVING DOCK A specially designed dock in which oil rigs and production platforms are built.

HYDROCARBONS Chemical substances, either liquids, solids or gases, which contain only carbon and hydrogen. North Sea oil and gas are mixtures of hydrocarbons.

JACK-UP RIG An oil rig whose legs are lowered onto the sea bed when it has been towed into position. The platform of the rig is then jacked up above the surface of the sea.

MUD During drilling a special mud, made up of a mixture of clay, water and certain chemicals, is pumped through the drilling string. The mud lubricates the drill bit, protects the drill pipe from the walls of the well, and carries fragments of rock to the surface which give information about the presence of oil or gas.

NATURAL GAS Gas found in areas such as the North Sea. It is made up of all hydrocarbons which turn from liquid to gas at a temperature of 60 degrees Fahrenheit under normal pressure.

PRODUCTION PLATFORM Once oil has been discovered by a drilling rig, a production platform is positioned over the well, and draws up the oil from beneath the sea bed. Platforms are made of either steel or concrete.

RECOVERABLE RESERVES The amount of oil or gas that can be pumped out of a field. Only a proportion of the oil or gas in a well can be recovered. This can vary between 10 and 80 per cent, but the average is between 20 and 40 per cent.

REFINERY When the crude oil has been pumped out of the ground it is piped to a refinery. Here it is split up into its various parts, called fractions, which are stored ready to be taken where they are needed.

SEMI-SUBMERSIBLE RIG A rig supported on large pontoons submerged in the sea which keep it stable in the water. This type of rig is suitable for use in deep waters.

STRING The pipe which carries the drilling bit down the oil well. It is made up of lengths of pipe screwed together, each about 30 feet long.

TERMINAL When oil or gas is pumped ashore it is stored in a terminal, before being refined.

TOWN GAS Before the full flow of natural gas from the North Sea, gas was made in gasworks from coal, then stored and pumped to the towns and cities to be used.

WHITE PAPER When the Government wants to present a bill in Parliament, it first of all sets out its ideas in a series of papers, the last and most definite of which is a White Paper.

Index

Picture Credits

The author and publishers wish to thank all those who have kindly given permission for their illustrations to appear on the following pages:
BP, 8, 17, 19, 20, 21, 31, 35, 42, 46, 48, 49, 54, 56, 57, 66, 68, 86; Conoco, 10, 24, 37, 72; Mobil, 4, 26, 34 (left), 36, 43, 52, 53; Shell, *frontispiece*, 12, 13, 14, 15, 32, 39, 45; Schlumberger, 33, 38; Keystone Press, 9, 75, 76, 82; Camera Press, 63; MacAlpine, 51; the Nature Conservancy Council, 83; the RSPCA 85.